BLOCKCHAIN OR DIE

Praise for *Blockchain or Die*

The rapid and far-reaching changes in digital technology require us to be au fait with and to keep abreast of this dynamic field. *Blockchain or Die* is a succinct, yet comprehensive, introduction to the world of cryptocurrency and blockchain technology, which are transforming how businesses operate in the twenty-first century, and which we anticipate will soon affect how even the man in the street conducts his personal affairs.

Eric's book is therefore a "one-stop shop" for anyone, regardless of their level of interest, knowledge, or involvement in the field, who wants to gain a basic understanding of these digital currencies. It is timely and relevant. Importantly, it is also a poignant reminder that "the future is now." I am therefore thrilled to recommend *Blockchain or Die* as a worthy addition to any collection of books.

> —*Her Excellency Audrey Marks*
> *Ambassador of Jamaica to the United States*

Blockchain or Die is an excellent book, and the secret sauce is the "Success Strategy Action Items" at the end of each chapter. There are many books and articles on cryptocurrencies and blockchains, but Eric uses the action items to really engage readers. Eric uses the same educational interactive experience in *Blockchain or Die* as his blockchain trainings.

As the Executive Director of the Government Blockchain Association, I recommend reading *Blockchain or Die* and following the recommended action items to start your blockchain experience.

> —*Gerard Dache, Executive Director*
> *Government Blockchain Association*

Blockchain or Die is an informative combination of business, technology, and law written in plain language. As the Managing Partner for the Cogent Law Group, I have read my fair share of blockchain content, and *Blockchain or Die* gives the reader the core legal information they need to understand cryptocurrency and blockchain regulations. The best part is Eric gives readers the resources to educate themselves on this topic. Once you read *Blockchain or Die*, especially the chapter on cryptocurrency regulations, you will know the basics and how to stay informed.

—*Thomas Goldstein, Esq.*
Managing Partner
The Cogent Law Group

After reading *Blockchain or Die*, I realized blockchain technology would forever change the business and higher education landscape. *Blockchain or Die* is so compelling that Barber Scotia College is planning to train our students on blockchain technology and using the blockchain as a component of our campus of the future. *Blockchain or Die* will be a cornerstone of the blockchain education for our students.

—*Dr. Melvin I Douglass, President*
Barber-Scotia College

Blockchain or Die is not only an excellent book; it's an excellent resource. Eric's book gave me the basics about cryptocurrencies and blockchain technology but it also gave me the next steps to learn more and actually use the technology. Eric also gives the reader the incentive to learn more about blockchain technology by showing how the technology can be used in many areas. As the Founder and Executive Director of the Made Man, our organization is an advocate for economic development and social justice and *Blockchain or Die* showed me how the Made Man can use blockchain technology to further our global mission. *Blockchain or Die* can do the same for your organization as well.

—*Dr. Ky Dele, M.A, M.S. ED*
Founder & Executive Director
The Made Man Foundation

BLOCKCHAIN OR DIE

LEARN TO PROFIT FROM
CRYPTOCURRENCIES AND BLOCKCHAINS

ERIC GUTHRIE, ESQ.

Better ME Better WE Publishing
ARLINGTON, VA

Better ME Better WE Publishing
PO Box 9850
Arlington, VA, 22219, USA
Ph: 1-202 709 9219 or online at www.BetterMEBetterWE.com

Ordering Information: Special discounts are available on quantity purchases by corporations, associations, and others. For details or general information about our other products and services, contact our Customer Care Department at the address above.

Cover design by Brian D. Johnson, Doran Designs, LLC
Editing and Interior Design by Tanya Brockett, HallagenInk.com

Blockchain or Die/ Eric Guthrie, Esq. —1st ed.
ISBN: 978-0-9979332-2-2 (Hardcover)
ISBN: 978-0-9979332-3-9 (Paperback)

Blockchain or Die *is dedicated to my parents, Dr. Milton Guthrie and Joyce Babb-Guthrie. Although this book is about the future of money using cryptocurrencies and blockchain technology, during our childhood our parents taught us (my siblings and me) about the value of money. Our parents taught us by example by saving enough money for our education and for their retirement by living within their means and making wise investment decisions. When my sister and I, Dr. Tracey Guthrie, started our first paper route, our parents taught us to put aside 20% for our own savings account. Thankfully, my parents are comfortably retired because their hard work and wise investments prepared them for a retirement where the lack of money is not a concern.*

Although my investment strategy contains far more risk than my parent's investment strategy, thanks to their financial wisdom, the majority of my investments are in stable and secure investment portfolios. The funding for my speculative investments, including cryptocurrencies, is based on what I can afford to lose.

I want to publicly thank my parents for instilling the importance of education, sound fiscal management, and hard work, but most importantly, I thank them for the undying love and support they gave me everyday of my life.

Table of Contents

CHAPTER 1

CHAPTER 2

Top Ten Cryptocurrencies by Market Capitalization

CHAPTER 3

CHAPTER 4

CHAPTER 5

CHAPTER 6

Blockchain Business and Government Applications

CHAPTER 7

CHAPTER 8

CHAPTER 9

Ten Ways to Make Money with Blockchain
Technology Today 201

CHAPTER 10

Cryptocurrencies, Blockchain Technology, and Positive Real-Life Impact

Foreword

At first glance, having an English major and theologian to write a foreword for a book on cryptocurrency and blockchains might not seem like a logical fit. However, the logic is in the curiosity that I possess about these topics and the insatiable desire I have to study, plan, and prepare for future occurrences.

Working with Eric and reading *Blockchain or Die* has been revolutionary for me. Before this book, the two concepts mentioned above were above my comprehension even though I began finding out about them in 2013. Further, this book not only gave me a working knowledge of these topics, but it also gave me something more valuable—an understanding of my previously-acquired knowledge. With that understanding now intact, my entire worldview and outlook on the future shifted, and I began to see cryptocurrency and blockchain systems already in place and some emerging.

As I explored this book more, I felt like new synapses were forming in my brain—like I had tapped into some yet untouched region that was waiting to be activated. Being the researcher that I am (and academic sleuth), I had to figure out more of this cryptic puzzle. Between the times when chapters came for my review, I did outside research. However, the same frustration arose when I explored or mined (a blockchain joke) the internet alone. I was gaining a lot of knowledge but not much understanding, and then I read *Blockchain or Die*.

I realized how Eric's carefully-crafted words—words that put complex concepts into layman's terms—created the sweet spot of understanding for the reader. *Blockchain or Die* is historic because it takes readers behind a curtain that explains the inner workings of this technological machine. Eric has done it. I understand these two topics, and I see many ways that they will continue to impact our future. I've even taken up studying them in the light of eschatology.

The book you are about to read is a product of many years of Eric's research and writing. As mentioned earlier, my worldview has changed after reading this book. I came away from *Blockchain or Die* with a broader view of the world *and* a sense of how blockchain technology makes the world a smaller place. Contradictions? Not really. I just see how much our world is changing and how much closer technology and finance are going to make us all. Eric has bridged a gap that needed to be bridged. This book is essential for anyone who wants to understand two major areas of the world as they are blossoming, and it is the book you must read if you want a glimpse of what the world will become once these two areas are full grown.

—Dr. Stephanie Freeman,
President of Metanoia University, Inc.

Preface
MY CRYPTOCURRENCY JOURNEY

The root problem with conventional currency is all the trust that's required to make it work. The central bank must be trusted not to debase the currency, but the history of fiat currencies is full of breaches of that trust. Banks must be trusted to hold our money and transfer it electronically, but they lend it out in waves of credit bubbles with barely a fraction in reserve. We have to trust them with our privacy, trust them not to let identity thieves drain our accounts.[1]

—Satoshi Nakamoto, Creator, Bitcoin

THE FUTURE OF MONEY is now. The future of business is now. We are in an age that is moving away from money printed with portraits of presidents to digital currencies founded by live innovators! We are in an age where stock exchanges are competing with global cryptocurrency exchanges. The future of money and business has arrived. If you are not currently investing in cryptocurrencies or blockchain technology, you are not investing in the future.

The above quote by Satoshi Nakamoto is the perfect one to start this book and this chapter. Of all of the Satoshi Nakamoto quotes, this is the one that made me believe in bitcoin. "Trust" is the key word that is repeated three times in this quote because the citizens have to trust governments and banks for the system to operate effectively. But sometimes that "trust" is only there because citizens do not have any other options. We have to use conventional currencies and conventional banks and trust them with our financial futures and our private information. But what if we had another option? What if governments were not the only currency creators? What if anyone could create a currency based on the day-to-day trust of the people because the currency transactions are managed and approved by those same people? The time has come; in fact, it has been here since 2008.

This is also the quote that made me question the realities of conventional currencies and start to research bitcoin. I never realized how much trust people put into governments and banks with a history of poor financial policy. During my lifetime, I witnessed: the Savings and Loan Crisis in 1980, Black Friday in 1987, and the Mortgage Crisis and Great Recession in 2008. All of these financial crises were started, in part, by breaches in trust by the government and large corporations that put profit before sound business principles.

If you are reading this book, you are open to learning more about a new system of currency, finance, and technology. This also means you are putting yourself in the position to profit from this new system. Learning this new system will expose you to a wealth of new information and challenge your beliefs in the current systems. Get ready to learn about the future. Get ready to profit from the future.

How I Learned About Bitcoin

As I started to research bitcoin, I realized that I did not fully understand fiat currencies, and ironically, I learned more about fiat currencies while researching bitcoin. I also learned bitcoin was used more as an investment than a cryptocurrency and I had a lot of experience in the investment world.

I have been investing in stocks for thirty years and in commodities and futures for twenty years. I was even a licensed Series 7 Financial Advisor for a few years. I have a passion for investing in stocks, and I frequently talk about stocks with coworkers, professional associates, and family members. Little did I know a new passion for investing was right around the corner. But before we talk about investing in cryptocurrencies or blockchain technology, please allow me to share my cryptocurrency story.

The Luncheon that Changed My Life

During a retirement luncheon in December 2016, a young coworker asked our luncheon group if we heard about bitcoin. I told him I recalled hearing about bitcoin a few years ago, but that was the extent of my knowledge. The young co-worker went into detail about bitcoin and cryptocurrencies. Fortunately, he was fairly knowledgeable and was able to answer our basic questions.

I immediately recognized bitcoin and cryptocurrencies as a groundbreaking technology and a rare investment opportunity. These are the kind of opportunities that make millionaires, even billionaires! After the luncheon, I spent the next three days reading every article I could find online on bitcoin and cryptocurrencies. I barely slept, and when I did sleep, I fell asleep reading cryptocurrency articles. I was not alone in my tireless research.

During my cryptocurrency discussions, I talked to people that stayed up for days reading everything they could about crypto-currencies. For example, the book *Digital Gold* tells Roger Ver's cryptocurrency origin story. As one of the early adopters of bitcoin, Roger was so captivated by the idea of a financial system that was not controlled by any governments that he read materials on bitcoin for days until he fell ill and called a friend to take him to the hospital.[2]

Fortunately, my intense research did not end in a hospital stay, but after three days of exhaustive research, I decided to invest in cryptocurrencies. I knew if I did not invest in cryptocurrencies, I would greatly regret my decision in 2028. I did not want to spend the next ten years living with regret that I did not invest in cryptocurrencies or incorporate blockchain technology into one of my business platforms. I joined blockchain organizations, eventually taking leadership roles, became a Blockchain Certified Consultant, became a Blockchain Certified Trainer, and trained blockchain certification courses domestically and internationally. But I wanted to share the wonders of this new technology far beyond the reach of my training, so I decided to write *Blockchain or Die*.

MY MOTIVATION TO WRITE *BLOCKCHAIN OR DIE*

During my 2016 and 2017 *Diversify or Die* book signings (that was my first published book), I talked about diversity business practices, marketing strategies, and using cryptocurrencies to diversify personal investments and business opportunities. In the over twenty book signings I conducted with a combined estimate of over a thousand people, five people knew about cryptocurrencies or blockchain technology. Four were in the IT (information technology) field, and one had a daughter who invested in bitcoin when it was less than a dollar. This mother of this very shrewd daughter knew her daughter made a lot of money for her retirement in bitcoin, but she herself did not understand or know how to invest in bitcoin.

Since many of the book-signing attendees were corporate executives and entrepreneurs, I was surprised more attendees did not know about bitcoin, cryptocurrencies, or blockchain technology. When I shared my cryptocurrency story at these events, attendees were so riveted, many took out their smart phones to take notes or visit websites I referenced. One gentleman sitting next to his wife literally said, "Baby, pass me pen and paper," and as the other attendees laughed, he started taking notes. They were laughing, but he was dead serious.

During the *Diversify or Die* book signing question and answer sessions, attendees always asked questions about my cryptocurrency investing, which I was happy to share. When attendees asked if *Diversify or Die* contained any information about cryptocurrencies, I had to tell them no. That was a sign I needed to write *Blockchain or Die.*

WHY THE NAME *BLOCKCHAIN OR DIE*?

Chances are you never heard of cryptocurrencies or blockchains or you heard of them but don't understand how they work. If most of the global population is at the same level of understanding, how does not using the blockchain lead to death? There are two reasons for the title *Blockchain or Die.*

First, after considering and testing many other titles, *Blockchain or Die* raised a level of interest and created a similar sense of urgency. But is there really a sense of urgency? That question takes us to the next answer.

Second, *Blockchain or Die* prepares the reader for future prosperity. One example I use in training, presentations, and conversations is the commercialization of the internet, which led to the dot-com boom and the dot-com bubble. Early dot-com investors and adopters made millions investing in dot-com companies, but then the dot-com bubble burst and investors that were still in the market suffered significant losses. This also presented profitable buying opportunities as many dot-com companies with a solid business plan and solid fundamentals were severely undervalued. Amazon, for example, returned massive profits for investors. History has a tendency to repeat itself, and after a massive cryptocurrency bubble (which will be covered in later chapters) in 2018, the cryptocurrency bubble burst. This presents another opportunity for investors. For example, at the time of publication, many cryptocurrencies have rebounded from their lowest prices in 2018.

Blockchain is revolutionizing business and industries the way the internet did in the 1990s. If businesses do not start researching the use of blockchains in their business or industry, they run the risk of missing the blockchain revolution while their competitors use blockchain to offer new services, increase security, and increase profits. This blockchain revolution may not happen for five years or even ten years, but it will happen, and when it does, will your business be ready? Your business can be ready for the blockchain revolution and thrive or ignore the blockchain revolution and eventually die. Thus, *"Blockchain or Die."*

I personally decided to not only join the blockchain business revolution, but to train and educate on the revolution that is already in progress.

WHO SHOULD READ *BLOCKCHAIN OR DIE*?

You are probably reading this book because you are interested in cryptocurrencies, blockchain technology, or both, but don't know where or how to start. That is a common concern as there are thousands of articles online and many books on cryptocurrencies and blockchain technology.

When I started to research bitcoin, cryptocurrencies, and later blockchain technology, I found hundreds of articles, blogs, and videos online that covered the general concept of cryptocurrencies or very specific aspects of cryptocurrencies. After one year of intense research, attending cryptocurrency and blockchain meetings, getting involved with organizations, conducting seminars, and designing blockchain courses, I noticed a gap between general concept articles and articles that addressed specific aspects of cryptocurrencies.

Specialized articles filled one more piece of the cryptocurrency puzzle, but it was almost impossible to find all of the pieces of the puzzle in one source. Whenever anyone asked me to send them cryptocurrency and blockchain articles, I would send over ten articles with an email explaining why each article was important. I also explained the list of articles only scratches the surface and there are many more articles and books on this topic, but I didn't want to overwhelm anyone by sending them hundreds of articles. This was a time consuming process and another sign that I needed to write *Blockchain or Die*.

WHAT YOU WILL LEARN FROM *BLOCKCHAIN OR DIE*?

Blockchain or Die introduces cryptocurrencies as an investment option and blockchain technology as a business solution to provide a solid foundation for your first or early exposure to this new technology. If your first cryptocurrency articles were very technical, you may have been discouraged by the technical cryptocurrency and blockchain terms. This book is written in very plain language and discusses the basics of cryptocurrencies and blockchains in an easy to understand manner, but also shows you the importance of understanding the technical terms associated with the new technology.

Blockchain or Die includes a brief history of bitcoin because a number of books already provide a very detailed history of bitcoin and other cryptocurrencies. These historical books tell the stories of the bitcoin leadership team and very early bitcoin adopters, including their personal lives and how their individual businesses directly impacted the development of bitcoin. The stories of the early supporters of bitcoin make for an interesting read. For example, Silk Road, one of the first websites that used bitcoin for payment, was an internet site on the dark web that allowed people to buy and sell illegal goods and services anonymously. The stories associated with Silk Road were very entertaining and educational as they exposed me to a world I had only previously experienced in movies and on TV. According to "BitcoinWiki.com," Silk Road was shut down by the Federal Bureau of Investigations (FBI) in October of 2013 and Silk Road 2.0 was shut down by the FBI and Europol in November of 2014.[3]

Blockchain or Die is designed to show the reader:
- Key cryptocurrency and blockchain terms.
- Digital currencies that preceded bitcoin.
- The digital currency "double spending" problem.
- About public keys and private keys.
- Basic bitcoin transactions.
- About the volatile nature of cryptocurrency investments.
- About initial coin offerings.

- How to identify cryptocurrencies by market share.
- The role of the government in cryptocurrency regulations.
- How smart contracts work.
- About Decentralized Autonomous Organizations.
- The importance of "user interface."
- The impact blockchain technology will have on many industries.
- Ways to make money using the blockchain.
- The impact blockchain technology has on people across the world.

If these terms are new to you, they will be discussed in detail many times throughout this book.

After you read this book, you will know where to find the following:

- Cryptocurrency wallets
- Cryptocurrency prices, volume, and market share
- Cryptocurrency exchanges
- Initial Coin Offerings
- Decentralized Applications (DApps)
- U.S. Government decisions that impact cryptocurrencies and blockchain

WHAT *BLOCKCHAIN OR DIE* DOES NOT COVER

Blockchain or Die does not address blockchain programming or coding; however, the importance of programming and coding should not be overlooked. Blockchain programming is programming for the future. Many employers are actively looking for blockchain programmers and they are willing to pay a higher salary for their talents and experience.

Many books and courses provide programming and coding education. For example, FreeCodeCamp, a donor-supported tax-exempt 501(c)(3) nonprofit organization, helps people learn to code for free. FreeCodeCamp has thousands of videos, articles, and interactive coding lessons and FreeCodeCamp study groups around the world.[4] GitHub.com is another easily accessible source for developers, aspiring developers, businesses, or anyone interested in following updates in blockchain technology. GitHub.com is a development platform and free resource, all you have to do is create an account and determine whether it can best suit your blockchain needs and aspirations.[5]

INTERESTED IN BLOCKCHAIN TECHNOLOGY BUT NOT CRYPTOCURRENCIES?

When I speak about the blockchain, many business owners and government officials state they are only interested in the blockchain, not cryptocurrencies. It is important to note cryptocurrencies and blockchain technology are usually interoperable. Blockchain technology makes bitcoin possible, but as covered in later chapters, blockchain technology has grown beyond bitcoin.

In many conversations, business owners and corporate executives personally shared their interest in blockchain technology, but not in cryptocurrencies. Their focus was on blockchain technology as an enterprise solution not cryptocurrencies as a speculative investment. But cryptocurrencies are more than speculative investment; they are also a means to finance the creation and operation of a blockchain business. Although blockchain technology was created to support bitcoin, it is helpful to review bitcoin to understand the origin and history of blockchain technology. Now the roles have reversed and cryptocurrencies are used to support blockchain business ventures. For that reason, *Blockchain or Die* should be read in its entirety.

EVOLVING WITH TECHNOLOGY

Like any worthwhile journey, the cryptocurrency and blockchain journey is one of continued evolution. Cryptocurrencies and blockchains are in their second decade of existence, and companies, legislatures, and regulators are learning ways to integrate both technologies into their infrastructures. The legislative and regulatory impact on cryptocurrencies will be especially important as the statements and actions by some government agencies have a profound impact on cryptocurrency prices and the pace of blockchain development.

Although bitcoin was the first cryptocurrency, there are other cryptocurrencies with business models that will evolve cryptocurrencies and blockchain technology. That is why the cryptocurrency and blockchain technology is a continuous journey, because the next bitcoin is one white paper, blog post, or conference presentation away.

[1]

AN INTRODUCTION TO CRYPTOCURRENCIES

"When bitcoin currency is converted from currency into cash, that interface has to remain under some regulatory safeguards. I think the fact that within the bitcoin universe an algorithm replaces the function of the government ...[that] is actually pretty cool."[6]

—Al Gore, Former Vice President of the United States

AL GORE'S QUOTE recognizes one of the main components of bitcoin and cryptocurrencies: the algorithm that replaces the government function enabling cryptocurrencies to operate in a decentralized environment. Bitcoin received most of the attention as it was the first successful cryptocurrency, but it was not the first cryptocurrency that attempted to use this type of algorithm in a decentralized environment. An analysis of cryptocurrencies that preceded bitcoin is important for a number of reasons.

1. The history shows cryptocurrencies are not a "fad" but an important evolution in finance and business.

2. The history of cryptocurrencies will help you understand the core components of bitcoin and other cryptocurrencies.

3. The history shows cryptocurrencies and blockchain technology are highly interoperable as the bitcoin blockchain made bitcoin possible and successful.

The next section briefly discusses cryptocurrencies that preceded bitcoin.

WERE CRYPTOCURRENCIES PREDICTED?

Except for the Euro (the currency for nineteen European member countries), every country has it's own currency known as a "fiat currency."[7] "Fiat" is Latin for "let it be done." Fiat currencies are not backed by any commodities and are solely based on the full faith and credit of the government. Fiat currencies are primarily used in the country backed by the government, which in effect creates national economic borders. Although the use of technology to globally buy and sell goods and services has dissolved global borders, fiat currencies maintain economic borders. For example, when individuals and businesses travel or conduct business between countries, they are required to exchange fiat currencies (e.g., exchange U.S. dollars for Mexican pesos). Bitcoin, and other cryptocurrencies, are not fiat currencies as there is no conclusive evidence it was not created by a government and does not require any exchange to fiat currencies to operate as a currency.

Is bitcoin, or another cryptocurrency, a currency the world needs? Is this the right time for a world currency? Although bitcoin is used as a currency, a world currency was theorized thirty years before bitcoin came into existence. *The Economist* article, "Get Ready for the Phoenix," predicted:

THIRTY years from now, Americans, Japanese, Europeans, and people in many other rich countries, and some relatively poor ones will probably be paying for their shopping with the same currency. Prices will be quoted not in dollars, yen, or D-marks, but in let's say, the phoenix. The phoenix will be favoured by companies and shoppers because it will be more convenient than today's national currencies, which by then will seem a quaint cause of much disruption to economic life in the last twentieth century. (Emphasis not added.)[8]

Aside from the name, "The Phoenix," this part of the article generally describes bitcoin. "Get Ready for the Phoenix," was published on January 9, 1988. Yes, 1988! This article predicted the widespread use of a global currency more than thirty years ago. Although the article does not specifically reference the phrase "digital currencies," the prediction seems to line up with the creation and use of bitcoin and other cryptocurrencies. Even the imagery of the artwork as the phoenix rises from the burning ashes of major fiat currencies is symbolic, as cryptocurrencies have started to replace the use of fiat currencies in some countries. This article discussed digital currency in theory, but even in the 1980s, innovators started to experiment with the idea of a digital currency.

BITCOIN WAS NOT THE FIRST CRYPTOCURRENCY

Two decades before bitcoin, a number of digital innovators attempted to create a digital currency. It appears Satoshi Nakamoto, the creator of bitcoin, used the lessons learned from the previous digital currency failures to create a successful digital currency. The most promising digital currencies that preceded bitcoin, which would later be called "cryptocurrencies," follow in the order of their creation.

DIGICASH

DigiCash, created by David Chaum in the late 1980s, was electronic cash that used public and private key cryptography. The DigiCash technology is very similar to the public and private keys used by bitcoin and other cryptocurrencies. In 1998, Digi-Cash ran out of money, declared bankruptcy, and was sold in 2002.[9]

HASHCASH

HashCash, created by Adam Back in 1997, used the proof of work algorithm to limit spam email and prevent distributed denial of service (DDOS) attacks. HashCash did not succeed because in 1997 computer processing was not fast enough to handle the processing rate to adequately process the HashCash protocols.[10]

BIT GOLD

Bit Gold, theorized by Nick Szabo in 1998, used a proof of work system, very similar to bitcoin, and conceptualized the movement from a centralized to decentralized currency. According to CoinCentral.com, Bit Gold was never coded, so it never moved from theory to practice.[11]

One "person" finally created a decentralized cryptocurrency: Satoshi Nakamoto. Nakamoto combined components from each digital currency that contributed to bitcoin's success. It combined:

- The public and private key technology from DigiCash,
- The proof of work algorithm from HashCash, and
- The proof of work system and decentralized component from Bit Gold.

But the blockchain is the main component that makes bitcoin and other cryptocurrencies possible (as covered later in the book). From that perspective, Nakamoto used the blockchain to create the first operable cryptocurrency.

WHO CREATED BITCOIN?

Everyone loves a good mystery. History is full of larger than life characters whose historic and groundbreaking actions have spawned articles, books, TV shows, and movies. Now there is a new global mystery: "Who is Satoshi Nakamoto?"

On the record, Satoshi Nakamoto is the inventor of bitcoin. As you will see later in this chapter, Satoshi published the Bitcoin White Paper, but no one really knows the true identity of Satoshi Nakamoto. The name "Satoshi Nakamoto" indicates he is a Japanese male; however, expert analysis of his early writings on the Bitcoin.org blog raises questions about his Japanese ancestry. There is speculation that Satoshi Nakamoto is not a person but a group of people that created an identity to start the bitcoin phenomenon.[12] There is also speculation that Satoshi Nakamoto is actually Nick Szabo or Hal Finley, both of whom worked with Satoshi from the very start of bitcoin.[13]

Blockchain or Die does not analyze the identity of Satoshi Nakamoto, as a number of books already provide a detailed analysis of the origins and identity. As an introduction to bitcoin, it is important to know the Satoshi Nakamoto story, but in the spirit of bitcoin, the fact that Satoshi Nakamoto may not be a real person should not prevent anyone from considering bitcoin, or cryptocurrencies, as potential investments, or blockchain technology as a business platform. When Satoshi Nakamoto left the bitcoin community in December 2010, he left the decentralized bitcoin protocols in the hands of the bitcoin community, which are still in operation to this day.

As discussed in the analysis of the Bitcoin White Paper abstract later in this chapter, "transparency" and "trust" are critical and integral components of the platform. If that is the case, the identity of Satoshi Nakamoto does not matter. The transparency and trustworthiness of the bitcoin protocol means the trust is in the decentralized and open source technology, not a centralized organization or government.

The following 2016 conversation with a colleague shows the way many people feel about bitcoin and cryptocurrencies.

"Do you invest in bitcoin?" I asked.

"No. I can't invest in anything when I cannot identify who created it, and if I cannot identify them, I cannot trust them."

"The trust and transparency of the bitcoin protocol is the reason to invest, not the person who created the protocol," I explained. "Satoshi Nakomoto handed over the bitcoin protocol to the bitcoin community before he vanished and the protocol is the magic behind bitcoin, not Satoshi."

"I still don't trust bitcoin."

"Have you heard of any other cryptocurrencies, for example, Ethereum, Litecoin, and Ripple? The creators of these coins are very visible inside and outside the cryptocurrency community."

"I don't know of those cryptocurrencies," he replied.

"Would you consider investing in these cryptocurrencies since everyone knows the creators of these cryptocurrencies?"

"Probably not," he said.

The lesson here is that transparency and trust are critical forward thinking components, but any level of involvement in cryptocurrencies or blockchain technology requires the understanding that we are moving away from a centuries old centralized financial system towards a transparent and trust based mathematical financial system. The reality is, like my colleague, most people are not ready to invest in cryptocurrencies. But the first bitcoin adopters offered their time, money, or expertise because of the 2008 Bitcoin White Paper.

THE WHITE PAPER THAT FOREVER CHANGED MONEY AND TECHNOLOGY

On October 31, 2008, Satoshi Nakamoto sent out a white paper titled: *Bitcoin: A Peer-to-Peer Electronic Cash System* (hereafter the "Bitcoin White Paper").[14] The Bitcoin White Paper changed the landscape of technology and finance. Satoshi only sent the white paper out to people in his network, but once he sent it out, his idea started to spread. While some recipients of Satoshi's email did not support his concept of a peer-to-peer electronic cash system, other recipients wholeheartedly supported Satoshi's idea and provided their skills to turn the idea of bitcoin into a reality.[15]

Satoshi's groundbreaking white paper was the breakthrough needed to overcome the hurdles to create a successful digital currency and set the stage for the next decade of development of bitcoin, cryptocurrencies, and blockchain technology. What is important about the white paper is not only the content, but also what is missing.

The Bitcoin White Paper does not talk about profit, market dominance, or any nation building, which were some of the reasons behind creating a decentralized digital currency. Bitcoin solved a problem that was once deemed unsolvable. Many papers and books have thoroughly analyzed the Bitcoin White Paper, bitcoin's history, and the events that made present day cryptocurrencies. *Blockchain or Die* does not seek to repeat the same

stories. Since this an introduction to bitcoin and cryptocurrencies, the Bitcoin White Paper "Abstract" is an excellent first step
to understanding bitcoin and cryptocurrencies.

AN INTRODUCTION TO THE BITCOIN WHITE PAPER

Generally, a white paper is a document that explains a complex
issue or proposes a solution to a problem. Each white paper starts
with an abstract, which is essentially a summary of the white paper, briefly explaining the issue or solution in the white paper.
The Bitcoin White Paper abstract follows:

> A purely **peer-to-peer** version of **electronic cash** would al-
> low online payments to be sent directly from one party to an-
> other without going through a financial institution. **Digital
> signatures** provide part of the solution, but the main benefits
> are lost if a trusted third party is still required to prevent **dou-
> ble-spending**. We propose a solution to the double-spending
> problem using a **peer-to-peer** network. The network
> timestamps transactions by hashing them into an ongoing chain
> of **hash-based proof-of-work, forming a record** that
> cannot be changed without redoing the proof-of-work. The
> longest chain not only serves as proof of the sequence of events
> witnessed, but proof that it came from the largest pool of CPU
> power. As long as a majority of CPU power is controlled by
> nodes that are not cooperating to attack the network, they'll
> generate the longest chain and outpace attackers. The network
> itself requires minimal structure. Messages are broadcast on a
> best effort basis, and nodes can leave and rejoin the network at
> will, accepting the longest proof-of-work chain as proof of what
> happened while they were gone.[16]

The Bitcoin White Paper abstract sets the stage for the rest of
the Bitcoin White Paper and introduces a number of terms creat-

ed for this new technology. In order to fully understand the foundational principles of bitcoin as presented in the Bitcoin White Paper, and cryptocurrencies as a whole, it is important to understand this new terminology. This book provides the definitions for the terms in the white paper. The bolded words in the abstract are important for understanding bitcoin and cryptocurrencies. A brief definition of the terms follows and the Glossary contains a more detailed version. The definitions of the key terms, in the order of appearance in the Bitcoin White Paper, follow.

Peer to Peer/P2P—A connection between two or more computers that allows them to directly share information, files, or other data.[17]

Electronic Cash/Cryptocurrency—Electronic money that uses technology to control how and when it is created and lets users directly exchange it between themselves, similar to cash.[18]

Digital Signature—Permission and proof done through a computer that an authorized person has agreed to something and generates a verification code that proves a transaction took place. Digital signatures are used by cryptocurrency systems to allow the owner to send and receive money.[19]

Double Spending—A form of deceit using digital money where the same money is promised to two parties but only delivered to one. If completed successfully, one of the two recipients will receive worthless money.[20]

Hash—A computer program that takes information and irreversibly turns it into a series of letters and numbers of a certain length. Bitcoin uses the SHA-256 hash algorithm (secure hash algorithm) to generate verifiably "random" numbers in a way that requires a predictable amount of CPU effort.[21] (If you want to see the SHA-256 algorithm, https://www.movable-type.co.uk/scripts/sha256.html has a SHA-256 hash "converter" that converts any message into a SHA-256 hash. For example, the SHA-256 hash message of "Eric Guthrie" is e461d5bba6eb11b6372d5ba9c54fe882f5d6f170483411018813a 7c4225743d2).

Hashing—The actual work done by the central processing unit (CPU) to confirm the electronic transactions (bitcoin transactions). Consider hashing like a transparent electronic audit where all transactions have to be confirmed by the peer-to-peer network before the transaction can be finalized.[22]

Forming a Record/Ledger—A book or other collection of records in which a person, business, or other group records how much money it receives and spends.

Although the word "ledger" is not actually used in the abstract, the phrase "forming a record" is used and is an extremely important reference serves as the basis for many of the definitions that follow.[23]

Proof of Work—A process for achieving consensus and building on a digital record on the blockchain. Proof of work users compete with their computers to solve a tough math problem. The first computer (defined later as a "node") to solve the problem is allowed to create new blocks and record information that earns them a reward in digital currency plus fees paid for each transaction.[24]

Nodes—Any computer, phone, or any other computing device that can receive, transmit, and/or contribute to the blockchain.[25]

CRYPTOCURRENCY TERMS YOU NEED TO KNOW

While the following terms are not in the Bitcoin White Paper abstract, they are very important to fully understanding the language of cryptocurrencies.

Address/Public Address—A string of 26-35 alphanumeric characters, beginning with the number 1 or 3, that identifies a cryptocurrency wallet. It is used as a way to safely receive cryptocurrency.[26] The site www.Blockchain.com gives a sample bitcoin address: 1BoatSLRHtKNngkdXEeobR76b53LETtpyT.[27]

Altcoin (or "alternate coin")—Any cryptocurrency except for bitcoin. "Altcoin" is a combination of two words: "alternative bitcoin" or "alternative coin." Examples include: Ripple, Steem, Monero, etc.[28]

ASIC/ASIC Miner (An application-specific integrated circuit)—"Integrated circuit" is just a computer chip. "Application-specific" means it was built for one specific purpose or computer application rather than a general-purpose application.[29]

Blockchain—A digital ledger used to prove a group of people came to an agreement about something. By using this ledger, every user is able to find out what amount of bitcoin has ever belonged to a particular address at a certain time period. The blockchain is supported by decentralized efforts of many miners. With this information, anyone can find out how much value belonged to each address at any point in history. More on blockchains in Chapter 5.[30]

Decentralized Autonomous Organizations (DAO)—A leaderless organization supported by a network of computers. To be decentralized, it must have no central location because it is running on a network of computers. And because there is no single leader and has its own rules to follow, it is autonomous, or self-governing. This positive quality, where many people keep do the work of maintaining their copy of the blockchain is known as "decentralized."

Fork/Hard Fork—A decision to make a permanent change to the technology used by a cryptocurrency. This change makes all new recordings (blocks) very different from the original blocks.[31]

Fork/Soft Fork—A change made to cryptocurrency technology creating a temporary split in the group of recordings (blockchain). This change creates all new, valid recordings (blocks) that are slightly different from the original blocks.[32]

Initial Coin Offering (ICO. Also "Initial Token Offering" or "ITO" and "Token Generation Event" or "TGE")—When a new cryptocurrency or token generally becomes available for public investment. ICOs are similar to Initial Public Offerings (IPOs) where a company raises money by selling public shares of their stock. More on ICOs in Chapter 4.[33]

Mining—The process of using computer power to solve a complex math problem presented by the crypto system, review and verify information, and create a new recording to be added to the blockchain. In mining, transactions are added to a public ledger of past transactions.[34]

Private Key—A string of random letters and numbers known only by the owner that allows them to spend their cryptocurrency. The private key is mathematically related to the cryptocurrency address, and is designed so the cryptocurrency address can be calculated from the private key, but importantly, the same cannot be done in reverse.[35]

Public Key—A code consisting of a string of letters and numbers that allows cryptocurrency to be received. Public keys are not considered as safe to use as public addresses. A "public key" isn't publicly visible until you've shared it or sent money out. Every bitcoin address has a public key, which coupled with the private key, ensures the security of the crypto- economy.[36]

Smart Contract (also self-executing contract, blockchain contract, or digital contract)—An agreement to exchange goods, services, or money that will automatically execute, without third party oversight, so long as established criteria are met. Smart contracts are electronic algorithms that automate the contract execution process in the blockchain. More on smart contracts in Chapter 5.[37]

Wallet—A collection of public and private keys, but may also refer to client software used to manage those keys and to make transactions on the bitcoin network. Wallets don't actually store the money; they lock away access. The only way to get access to the money is by providing a password, as defined earlier, more commonly referred to as a "key."[38]

As with any new language, it takes time to fully understand the exact meaning of essential terms. The intent of this section is to provide a basic understanding of cryptocurrency and blockchain language to understand their use in articles, blogs, books, or even conversation. Additional research and actual use of cryptocurrencies and blockchain technology is the best way to understand this new technology. Finally, these terms will be reintroduced throughout the book in areas where the term is used.

HOW DOES BITCOIN REALLY WORK?

Now that the Bitcoin White Paper and the key cryptocurrency definitions have been explained, we can analyze a bitcoin transaction as described in the Bitcoin White Paper.

New transactions are broadcast to all the nodes.

1. Each node collects new transactions into a block.

2. Each node works on finding a difficult proof-of-work for its block.

3. When the node finds a proof-of-work, it broadcasts the block to all the nodes.

4. Nodes accept the block only if all of the transactions in it are valid and not already spent.

5. Nodes express their acceptance of the block by working on creating the next block in the chain, using the hash of the accepted block as the previous hash.

Sounds like a foreign language? Hopefully the definitions earlier in the chapter will help explain the new terms. A real world example in the same numerical sequence of the same transaction follows. The key terms defined earlier in the chapter are bolded for reference.

Eve wants to transfer her bitcoin to another person, so she initiates a transfer from her **wallet** by signing off with her **private key**. Eve's transaction is broadcast on the **blockchain** network to the **miners**.

1. Each node collects Eve's transactions into a block.

2. The **miners** in the network start **hashing** to make sure Eve's **wallet** has sufficient bitcoin and her transaction is added to the block of recent transactions.

3. The **nodes** start the race to be the first one to have their transactions added to the blockchain.

4. The **node** that is the first to complete the task accepts the block and receives a predetermined amount of bitcoin for being the first to complete the computation.

5. Eve's transaction is complete and the next set of transactions are broadcast and submitted to the **blockchain** network.[39]

This sounds complicated but it is essentially the role of centralized organizations (e.g., banks and credit card companies) to verify transactions from one source to another. There are three major differences:

1. The current financial system is based on fiat currencies not cryptocurrencies.
2. The current financial system is centralized, so only centralized organizations and paying customers can see their transactions.
3. Centralized systems charge the account holder to verify transactions; in the blockchain, the miners are paid for the hashing.

These are important distinctions as digital technology and decentralized digital currencies created a new challenge: double spending.

THE SOLUTION TO COPYING DIGITAL MONEY

Anything in digital form can be easily copied. For example, copying an actual book requires actually copying the book as a physical object. Copying physical objects requires an investment of time and money. After the copying is completed, sharing the book requires more time and money. However, a digital version of any book can easily be copied and distributed within seconds. The same principle applies to digital currencies.

Digital currencies can be easily duplicated and distributed. Only the original version of the digital currency has any value and the recipient does not know if they are receiving the original digital currency or the duplicate currency. If a duplicate currency was created and distributed, at some point multiple recipients will attempt to spend the same duplicated digital currency and learn their digital currency is worthless. This is called "double spending." Why is double spending a problem?

One of the main reasons double spending exists is because all of the transactions are kept on a central private ledger. Spenders do not know if the actual digital currency they received was spent in a prior transaction or if they are in possession of a worthless copy of a currency. Another factor that enables double spending is that centralized transactions are verified based on the latest transaction. But with bitcoin, miners verify blockchain transactions from the first transaction to the most recent transaction. Bitcoin solved both of those problems by creating a distributed ledger called the "blockchain," which will be discussed in detail in Chapter 5.[40]

ENCRYPTION AND DECRYPTION EXPLAINED

Encryption and decryption are important concepts as they provide the security for bitcoin and blockchain transactions. Bitcoin would not function without encryption and decryption. In bitcoin terms, the public key encrypts the data and the private key decrypts the data.

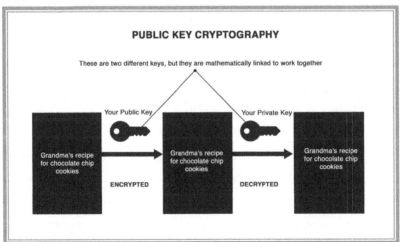

Figure 1.1 Public key cryptography

The diagram in figure 1.1 shows the public key as a "key," but the public key, also known as a bitcoin address, is a string of thirty-four letters and numbers. The private key is generally a string of sixty-four letters and numbers. The thirty-four-character public key and a sixty-four-character private key security feature make hacking the bitcoin protocol "computationally impractical."[41] "The Technical Appendix" in the book *Digital Gold* provides an excellent example of the public key and private key (shown below).

The bitcoin address example:

16R5PtokaUnXXXjQe4Hg5jZrfW69fNpAtf

The private key for this bitcoin address:

5JJ5rLKjyMmSxhauoa334cdZNCCoVEw6oLfMpfL8H1w9pyDoPMf3

The *Digital Gold* Technical Appendix explains each bitcoin address only has one private key and the person with the private key is authorized to complete a

> *Private Key—A string of random letters and numbers known only by the owner that allows them to spend their cryptocurrency.*

transaction. It's important to understand the functions of the public and private key in cryptocurrency wallets and how they increase security to protect from hacking threats.[42]

How Should You Store Your Cryptocurrencies

Most people use physical wallets for easy access to their identification, money, credit cards, and debit cards. Cryptocurrency wallets are used to securely access cryptocurrencies. The core feature of the cryptocurrency wallet is the private key, which enables the owner of the wallet to transfer cryptocurrencies in and out of the wallet. There are many types of wallets to store cryptocurrencies. This section will introduce and explain of each type of cryptocurrency wallet.

Hardware wallets are stand-alone devices, such as a USB or external hard drive, which are not connected to the internet or blockchain until connected through a device. Hardware wallets store all of the cryptocurrencies off-line, which provides a great deal of security for your cryptocurrencies. Never lose your hardware wallet, as it is the only way to access your coins.[43]

> *Wallet—A collection of public and private keys, but may also refer to client software used to manage those keys and to make transactions on the bitcoin network.*

Paper wallets, also known as "cold storage," are similar to hardware wallets in that they are not connected to the internet. Usually, paper wallets are printouts of a user's public and private key; however, paper wallets can also be a printout of the software that stores public and private keys. Since paper wallets are not connected to the internet, once users download and print their paper wallet, they should not save the paper wallet on their computer and they should clear their cache after printing the paper wallet. Clearing the cache will prevent hackers from obtaining your paper wallet or printing another version of your paper wallet. Paper wallets with one public key and one private key will only allow transfer of cryptocurrencies once. Finally, similar to the hardware wallet, be sure not to lose your paper wallet as it is the only place the coins are stored.[44]

For example, Cryptoart is a company that makes artistic paper wallets. Cryptoart stores cryptocurrencies by combining art, technology, and cryptocurrencies. After the cryptocurrency is stored, users can use the security code on the back of the Cryptoart to retrieve the cryptocurrencies from an online cryptocurrency exchange. The Cryptoart gives the instructions on using the art to store and retrieve cryptocurrencies.

Desktop wallets are downloaded and installed on personal computers or laptops. Once installed, cryptocurrencies can only be accessed from the same device. Since desktop wallets have greater exposure to the internet, they can be compromised if the personal laptop or computer is exposed to a virus.[45]

Online wallets are very easy to access as they can be accessed from any device from any location; all you need is an internet connection, a user name, and a password. A third party controls online wallets and they are prime targets for hackers since they may store millions of accounts.[46]

Mobile wallets are run by smart phone apps. As a third party controls them, mobile wallets have the same vulnerabilities as the desktop or online wallets.[47]

MULTI-CRYPTOCURRENCY STORAGE OPTIONS

Multi-cryptocurrency wallets enable use of several currencies in the same wallet. Multi-cryptocurrency wallets are helpful for everyday purchases or buying and holding digital currency for investment. Multi-cryptocurrency wallets allow access to digital wallets from any device connected to the internet.[48]

Which wallets suit your cryptocurrency needs? The most common wallet requirements include:
- Security
- Accessibility

- Ease of use/User interface
- Access to Wi-Fi

Users should determine their personal wallet requirements and then choose the most suitable wallet.

WHAT YOU NEED TO KNOW ABOUT HACKING

As use of technology has increased, so has hacking. Hacking is one of the most common concerns I hear in cryptocurrency conversations. Many people expressed the fear of hacking as a reason not to buy or invest in cryptocurrencies. The problem is many established and commonly used companies and government agencies face hacking threats everyday, and the vast majority of people continue to use their products and services.

The reality is hacking threats are everywhere, everyday. Many industries face hacking threats, and moreover, have been actually hacked. For example, NBCnews.com reports Target Corporation was the victim of one of the largest U.S. retailer data breaches. As a result, Target agreed to pay forty-seven states and the District of Columbia $47 million to settle claims. But Target is only the tip of the hacking iceberg. A Money.CNN.com article, "The Hacks that Left Us Exposed in 2017," lists a number of companies that were seriously hacked, including:

- Yahoo—Three billion accounts were hacked
- FedEx—Paid $300 million in ransom in a hack by Notpetya
- Uber—Financial data in 57 million accounts were hacked

Target, Yahoo, FedEx, and Uber are established billion dollar companies using established and centralized information technology (IT) security systems, and all of them have been hacked.[49]

Satoshi anticipated the hacking issue and designed bitcoin to greatly reduce the incentive to hack the bitcoin protocol. As stated in the Bitcoin White Paper:

"The [incentives] may help encourage nodes to stay honest. If a greedy attacker is able to assemble more CPU power than all of the honest nodes, he would have to choose between using it to defraud people by stealing back his payments, or using it to generate new coins. He ought to find it more profitable to play by the rules, such rules that favor him with more new coins than everyone else combined, than to undermine the system and the validity of his own wealth."[50]

According to Hacker-moon.com, in an incident commonly called the value overflow incident, hackers created 184,467 billion bitcoin out of nothing. Within hours of the hack, Satoshi responded quickly

> *Fork/Hard Fork—A decision to make a permanent change to the technology used by a cryptocurrency. This change makes all new recordings (blocks) very different from the original blocks.*

by creating a hard fork to remove the 184.467 billion bitcoin and preserve the future of bitcoin.[51]

When discussing hacking, it's important to distinguish hacking bitcoin from hacking cryptocurrency exchanges that store bitcoin. Cryptocurrency exchanges, trading platforms, and wallet services have been hacked resulting in a loss of customer's bitcoin. According to Artstechnica.com, the following cryptocurrency trading platforms were hacked resulting in a loss of bitcoin.

- MyBitcoin—August 2011: MyBitcoin was a popular wallet service that used to store bitcoin removed itself from the web claiming they were hacked.

- Linode—March 2012: Linode, a shared online web host, used to store bitcoin. The Linode web hosting had a vulnerability and hackers stole at least 43,703 bitcoin, which at the time was worth more than $200,000.

- Bitfloor—September 2012: Hackers stole 24,000 bitcoin, which was worth around $250,000. As Bitfloor did not have $250,000 to repay their customers, Bitfloor went out of business.

- Mt. Gox—February 2014: In the early days of bitcoin, Mt. Gox (Magic: The Gathering Online eXchange) was the leading bitcoin exchange to buy and sell bitcoin. It was hacked and lost 850,000 bitcoin worth an estimated $450,000 million. Mt. Gox went out of business.

- Bitstamp—January 2015: Bitstamp was hacked and lost an estimated 19,000 bitcoin worth an estimated $5 million. Bitstamp remained solvent and is still in operation as of the time of publication.

- Bitfinex—August 2016: Bitfinex was hacked and lost $77 million worth of bitcoin. As of the date of this publication, Bitfinex remains solvent and is still in operation.[52]

- Bithumb—June 2018: Bithumb was hacked and lost $31 million worth of bitcoin. Immediately after the hack, Bithumb confirmed it would use their reserves to pay back the victims of the hack.[53]

SUMMARY

Since the Bitcoin White Paper abstract served as the cornerstone of this chapter, it seems appropriate to end this chapter with a brief summary of the Bitcoin White Paper, and as stated in Satoshi Nakamoto's quote at the beginning of the chapter, the key word is "trust." Bitcoin innovatively combined existing technologies to create a system of trust to solve the double spending problem by using peer-to-peer networks and proof of work through nodes to create a virtually immutable public record of transactions. This decentralized system of transactions and public records creates an "honest" system. This is bitcoin, the start of the cryptocurrency and blockchain revolution.[54]

Although the Bitcoin White Paper was written in 2008, bitcoin and cryptocurrencies are still in their infancy. There are still many critical milestones for bitcoin and cryptocurrencies including: passing national and global security regulations, formal adoption by state and national governments, use as a payment by billion dollar businesses, and use as a currency by people around the world. Many of these concepts will be discussed in the following chapters.

SUCCESS STRATEGY ACTION ITEMS

1. Read Satoshi Nakamoto's Bitcoin White Paper *Bitcoin: A Peer-to-Peer Electronic Cash System* at least three times. The definitions in this chapter should help explain some of the key concepts in the white paper.
2. Use Decryptionary.com as one reference for information on cryptocurrencies. As many other websites provide information on cryptocurrencies, choose the online reference that best suits your learning style.
3. Create a bitcoin wallet. It's one thing to read about bitcoin and other cryptocurrencies; it's better to actually use them. You can go on to www.Bitcoin.org and create your own bitcoin wallet. Creating a bitcoin wallet is free and it will give you a better idea how to use bitcoin. If you are interested in other cryptocurrencies, you can create a wallet for your cryptocurrency of choice.

4. If you want to see the SHA-256 hash in action, use the https://www.movable-type.co.uk/scripts/sha256.html message converter.

5. If you want to learn more about the history of bitcoin and other cryptocurrencies, use the resources in the Appendix.

[2]

TOP TEN CRYPTOCURRENCIES BY MARKET CAPITALIZATION

> *"Thousands of crypto companies will be created and go public, but only a few will be massive successes."*[55]
>
> **—James Altucher, Best-selling Author,** Choose Yourself *and an editor at* The Altucher Report

ALTHOUGH JAMES ALTUCHER is not a household name, Mr. Altucher is an important figure in the cryptocurrency space and his words ring true with many cryptocurrency investors.

Chapter 1 introduced bitcoin and referenced other cryptocurrencies. This chapter expands *Blockchain or Die* beyond bitcoin and focuses on the top ten cryptocurrencies by market capitalization, which one could argue are already "massive successes."

"BLUE CHIP" CRYPTOCURRENCIES

Blue chip stocks are large billion dollar publicly traded companies that have been in existence for at least a decade and are market leaders in their industry. Blue chip stocks are usually considered low risk investments as many have very successful track records and the vast majority pay a dividend to their investors. This begs the question: Are there any "blue chip" cryptocurrencies? While blue chip cryptocurrencies do not exist, there are cryptocurrency market leaders with a successful track record. CoinMarketCap.com is an online resource to analyze cryptocurrencies.

COINMARKETCAP AS A RESOURCE

CoinMarketCap, a centralized resource for cryptocurrency information, tracks the market capitalization (defined later), price, volume, and circulating supply of all cryptocurrencies listed on their site. On August 9, 2018, there were 1,799 cryptocurrencies listed on CoinMarketCap.com with a total market capitalization of $231,159,489,538! By comparison, on January 20, 2019, the number of cryptocurrencies increased to 2,116 with a total market capitalization of $120,009,557,105.[56] This decrease in market capitalization will be discussed later.

Requirements for a cryptocurrency to be listed on CoinMarketCap.com include:

- Must be a cryptocurrency or a crypto token
- Must be on a public exchange with an application program interface (API) that reports the last traded price and the last 24-hour trading volume
- Must have a non-zero trading volume on at least one supported exchange so a price can be determined
- For market cap ranking, an accurate circulating supply figure is required[57]

WHAT IS THE DIFFERENCE BETWEEN A CRYPTOCURRENCY AND A CRYPTO TOKEN?

New cryptocurrency investors frequently ask about the difference between a "cryptocurrency" and a "crypto token." CoinMarketCap.com defines a cryptocurrency as a "digital medium of exchange using strong cryptography to secure financial transactions, control the creation of additional units, and verify the transfer of assets." CoinMarketCap.com defines a crypto token as "a digital unit designed with utility in mind, providing access and use of a larger crypto-economic system.[58]

"Crypto tokens do not have store of value on its own, but are made so software can be developed around it." Another way to explain these definitions is to compare Ether (ETH), as a cryptocurrency, and Binance Coin (BNB) as a crypto token built on the Ethereum platform. As a cryptocurrency, Ethereum has its own blockchain platform and its own currency, while Binance Coin uses the Ethereum blockchain platform to function.

The CoinMarketCap.com "Top 100 Tokens by Market Capitalization" list accurately illustrates the difference between cryptocurrencies and tokens.

According to an excerpt of the list in figure 2, all tokens except #1 are based on the Ethereum platform. Tether (#1 on this list) is the number one crypto token hosted on the Omni platform.[59]

Figure 2 Top ten tokens by market capitalization

BITCOIN SUPPLY AND DEMAND

Although bitcoin and cryptocurrencies are a new concept, the centuries old concept of supply and demand still applies. Twenty-one million bitcoin is the maximum number of bitcoin that will ever be created, so bitcoin is designed to increase in value. When it comes to bitcoin supply and demand, CoinMarketCap.com, provides the following definitions:

- Market Capitalization is one way to rank the relative size of a cryptocurrency and is calculated by multiplying the *Price* by the *Circulating Supply* (Price x Circulating Supply = Market Cap). For example, the Market Cap for bitcoin cash is calculated as follows: $130.40 x 17,582,925 = $2,292,861,428.

- Circulating Supply is the best approximation of the number of coins circulating in the market and in the general public's hands.
- Total Supply is the total amount of coins currently in existence (minus any coins that have been verifiably burned).
- Maximum Supply is the best approximation of the maximum amount of coins that will ever exist in the lifetime of the cryptocurrency. For example, Satoshi created twenty-one million bitcoin as a maximum supply.[60]

Circulating Supply is the best metric for determining market capitalization. Coins that are locked, reserved, or cannot be sold on the public market can't affect the price and are not allowed to affect the market capitalization. The method of using the Circulating Supply is analogous to the method of using public float—the number of stock shares held by public investors—to determine the market capitalization of companies in traditional investing.

As of June 12, 2019, figure 3 shows the top ten cryptocurrencies by market capitalization.[61]

Figure 4 shows the top ten cryptocurrencies by market capitalization as of January 9, 2019 (six months earlier than figure 3).

Although market capitalization is subject to change, the top ten cryptocurrencies have established themselves as consistent market leaders in the cryptocurrency space. As the next chart in figure 5 shows, the market capitalization has not changed much. In August 28, 2018, the top ten cryptocurrencies by market capitalization were:

#	NAME	MARKET CAP	PRICE
1.	Bitcoin	$144,683,199,615	$8,148.65
2.	Ethereum	$27,396,567,735	$257.33
3.	XRP (Ripple)	$17,027,601,016	$0.440361
4.	Litecoin	$8,342,138,256	$134.13
5.	Bitcoin Cash	$7,035,302,870	$398.48
6.	EOS	$5,876,766,250	$6.40
7.	Binance Coin	$4,881,874,908	$34.58
8.	Bitcoin SV	$3,385,063,692	$189.83
9.	Tether	$3,373,283,586	$1.01
10.	Stellar	$2,452,158,584	$0.126404

Figure 3 Top ten cryptocurrencies by market capitalization on June 12, 2019.

#	NAME	MARKET CAP	PRICE
1.	Bitcoin	$70,762,404,175	$4,050.43
2.	Ethereum	$15,746,707,219	$151.03
3.	XRP (Ripple)	$14,898,900,670	$0.365222
4.	Bitcoin Cash	$2,825,737,651	$160.96
5.	EOS	$2,545,625,158	$2.81
6.	Litecoin	$2,389,652,158	$39.87
7.	Stellar	$2,362,374,602	$0.123276
8.	Tether	$1,906,175,130	$1.02
9.	TRON	$1,738,132,471	$0.026081
10.	Bitcoin SV	$1,546,736,488	$88.11

Figure 4 Top ten cryptocurrencies by market capitalization on January 9, 2019.

#	NAME	MARKET CAP	PRICE
1.	Bitcoin	$144,683,199,615	$8,148.65
2.	Ethereum	$27,396,567,735	$257.33
3.	XRP (Ripple)	$17,027,601,016	$0.440361
4.	Bitcoin Cash	$8,342,138,256	$134.13
5.	EOS	$7,035,302,870	$398.48
6.	Stellar	$5,876,766,250	$6.40
7.	Litecoin	$4,881,874,908	$34.58
8.	Tether	$3,385,063,692	$189.83
9.	Cardano	$3,373,283,586	$1.01
10.	IOTA	$1,672,022,767	$0.601549

Figure 5 Top ten cryptocurrencies by market capitalization on August 28, 2018.

As the above charts show, Bitcoin, Ethereum, and Ripple retained their respective 1 – 3 positions in market capitalization in both tables. The 4 – 10 cryptocurrencies showed some changes in position, but over this time period as quoted in the chart: Bitcoin, Ethereum, XRP (Ripple), Bitcoin Cash, EOS, Stellar, Litecoin, and Tether have consistently remained in the top ten.[62]

In line with the quote that started this chapter, of the 2,235 cryptocurrencies as of June 12, 2019, the top ten cryptocurrencies on CoinMarketCap are successful and represents .00447% of the cryptocurrencies.

ANALYSIS OF TOP TEN CRYPTOCURRENCIES BY MARKET VALUE

The bitcoin early adopters took a massive risk when they invested in bitcoin even though previous attempts at cryptocurrencies failed. Since they failed, it is reasonable to assume the early investors in the cryptocurrencies before bitcoin lost some or all of their investment. Early adopters and investors invested in bitcoin when it was brand new and untested, but that has changed. Bitcoin and other cryptocurrencies now have successful track records and billions in market value. The analysis of the top ten cryptocurrencies by market value below includes the following information:

- CoinMarketCap.com information (name, price, chart, etc.)
- Currency Name and Symbol: Some cryptocurrency businesses name their currency after their business and some give their currency a different name. The analysis includes the symbol to help you identify the cryptocurrency quotes.
- Year Created: Shows the longevity of the cryptocurrency in the market.
- Creator(s): The success of the business is often based on the experience and track record of the founder or CEO. Knowing the creator's name provides the information to conduct additional research.
- Industry: Provides the specific industry in the cryptocurrency.
- General Information: Information about each cryptocurrency directly quoted from their website.
- All of the graphs were downloaded on or about July 11, 2019, from CoinMarketCap.com.

#1 Bitcoin

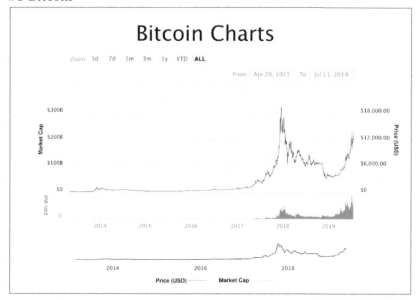

Currency Name: Bitcoin (BTC)

Year Created: 2009

Creator(s): Satoshi Nakamoto

Primary Industries: Currency

General Information:

"Bitcoin uses peer-to-peer technology to operate with no central authority or banks; managing transactions and the issuing of bitcoin is carried out collectively by the network. Bitcoin is open-source; its design is public; nobody owns or controls bitcoin and everyone can take part. Through many of its unique properties, bitcoin allows exciting uses that could not be covered by any previous payment system."[63]

Notice the market cap of bitcoin in the graph is a dotted line. The graphs in the rest of the chapter have multiple lines of different types (shown in the legend) because bitcoin is the standard by which all of the other cryptocurrencies are compared.

BITCOIN FORKS

As defined in Chapter 1, there are essentially two kinds of forks: hard forks and soft forks. According to Forkdrop.io, as of February 6, 2019, bitcoin has seventy-three active project/forks, of which fifteen are on CoinMarketCap.com.

According to CoinMarketCap.com, as the number four cryptocurrency by market cap at the time of publication, Litecoin is the most successful bitcoin fork. The next five most successful bitcoin forks by price per coin include:

- Bitcoin Cash
- Bitcoin SV
- BitcoinX
- Bitcoin Gold
- Bitcoin Diamond[64]

Many more bitcoin forks will be announced as bitcoin and its market evolves.

#2 Ethereum

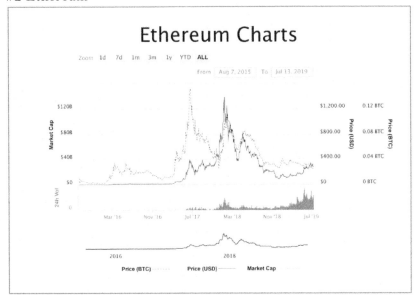

Currency Name: Ether (ETH)

Year Created: 2015

Creator(s): Vitalik Buterin

Primary Industries: Application Platform

General Information:

"Ethereum is a decentralized platform that runs smart contracts: applications that run exactly as programmed without any possibility of downtime, censorship, fraud, or third party interference. These apps run on a custom built blockchain, an enormously powerful shared global infrastructure that can move value around and represent the ownership of property. This enables developers to create markets, store registries of debts or promises, move funds in accordance with instructions given long in the past (like a will or a futures contract), and any many other things that have not been invented yet, all without a middleman or counterparty risk."[65]

#3 Ripple

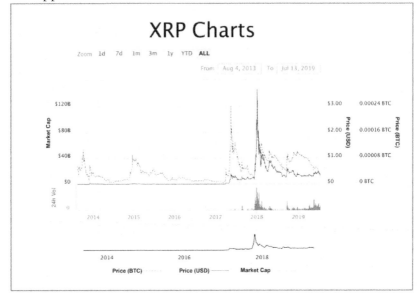

Currency Name: XRP
Year Created: 2016
Creator(s): Chris Larsen and Jeb McCaleb
Primary Industries: Payment and Exchange Network
General Information:

"In a world where three billion people are connected online, cars can drive themselves, and appliances can communicate, global payments are stuck in the disco era. Why? The payment structure was built before the internet with few updates. Ripple connects banks, payment providers, digital asset exchanges, and corporates via Ripplenet to provide one frictionless experience to send money globally. [Ripple is b]uilt on the most advanced blockchain technology that is scalable, secure, and interoperates different networks. [Ripple p]rovides optional access to the world's fastest and more scalable digital asset for payment." [66]

#4 Bitcoin Cash

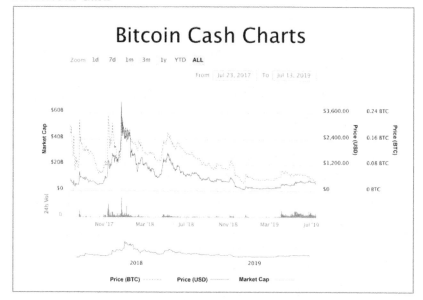

Currency Name: Bitcoin Cash (BTC)

Year Created: 2017

Creator(s): Amary Sechet and other supporters

Primary Industries: Currency

General Information:

"Bitcoin Cash, a fork of the bitcoin protocol, brings sound money to the world, fulfilling the original promise of Bitcoin as "Peer-to-Peer Electronic Cash." Merchants and users are empowered with low fees and reliable confirmations. The future shines brightly with unrestricted growth, global adoption, permissionless innovation, and decentralized development. "All are welcome to join the Bitcoin Cash community as we move forward in creating sound money accessible to the whole world."[67]

Unlike other cryptocurrencies on the top ten list, Bitcoin Cash is a hard fork of Bitcoin. A streamlined explanation on

BitcoinCash.org shows the differences between the Bitcoin and Bitcoin Cash as follows.

"1. New Name—Bitcoin Cash is more like 'cash' and easy to exchange with minimal or no fees.

2. Block Size Limit Increase—Increases of the block size limit to 8MB [megabytes].

3. Replay and Wipeout Protection—If and when Bitcoin Cash splits, there will be minimum disruptions.

4. New Transaction Type—Bitcoin Cash has introduced a new transaction type with additional benefits such as input value signing for improved hardware wallet security, and elimination of the quadratic hashing problem."

#5 Litecoin

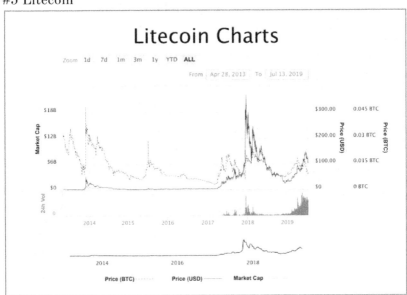

Currency Name: Litecoin (LTE)
Year Created: 2011
Creator(s): Charles (Charlie) Lee
Primary Industries: Currency

General Information:

"A decentralized online currency created in October 2011. Litecoin can be used to purchase services, such as website development, or to buy goods like jewelry or tea. Litecoin provides a safe and easy way for merchants to accept money as there are no fees to receive payments and no chargebacks. All transactions are recorded on a public ledger known as the blockchain so payments can be immediately verified. With Litecoin you can send money anywhere in the world to anyone instantly. The transaction fees are considerably less than those charged by major credit card companies, traditional bank transfers, and even other digital payment processors."[68]

#6 EOS

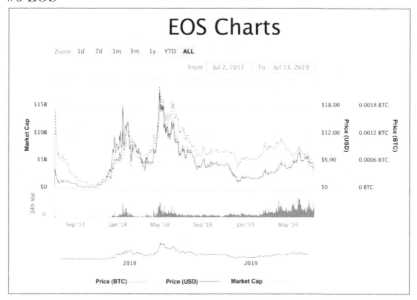

Currency Name: EOS (EOS)
Year Created: 2018
Creator(s): Daniel Larimer and Brendan Blumer
Primary Industries: Application Platform

General Information (According to the EOS.io White Paper v2 abstract):

"The EOS.IO software introduces a new blockchain architecture designed to enable vertical and horizontal scaling of decentralized applications. This is achieved by creating an operating system-like construct upon which applications can be built. The software provides accounts, authentication, databases, asynchronous communication, and the scheduling of applications across many of CPU cores or clusters. The resulting technology is a blockchain architecture that may ultimately scale to millions of transactions per second, eliminates user fees, and allows for quick and easy deployment and maintenance of decentralized applications, in the context of a governed blockchain."[69]

#7 Binance Coin

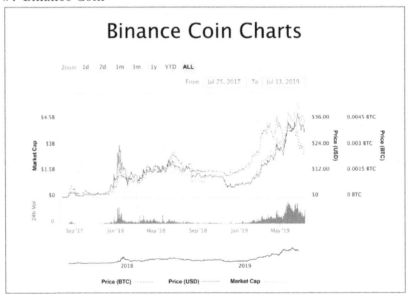

Currency Name: Binance Coin (BNB)
Year Created: 2017
Creator(s): Changpeng Zuao and Yi He
Primary Industries: Cryptocurrency Exchange

General Information:

"Binance is a blockchain and cryptocurrency exchange that provides the following services:

Academy—Blockchain and cryptoeducation
BCF—Blockchain charity foundation
Info—Cryptocurrency information platform
Labs—Incubator for top blockchain projects
Launchpad—Token launch platform
Research—Institutional-grade analysis and reports
Trust Wallet—Binance's official crypto wallet"[70]

#8 Bitcoin SV

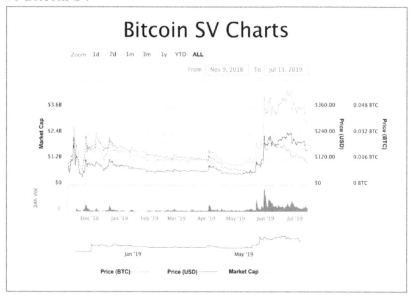

Currency Name: Bitcoin SV (BSV)
Year Created: 2008
Creator(s): Craig Wright
Primary Industries: Currency
General Information:

"Bitcoin SV is the original Bitcoin. It restores the original Bitcoin protocol, will keep it stable, and allow it to massively scale. Bitcoin SV will maintain the vision set out by Satoshi Nakamoto's white paper in 2008: *Bitcoin: A Peer-to-Peer Electronic Cash System.* "Reflecting its mission to fulfill the vision of Bitcoin, the project name represents the 'Satoshi Vision' or SV. Created at the request of leading BSV mining enterprise CoinGeek and other miners, Bitcoin SV is intended to provide a clear choice for miners and allow businesses to build applications and websites on it reliably."[71]

#9 Tether

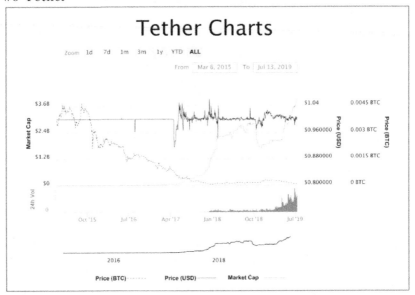

Currency Name: Tether (USDT)

Year Created: 2012

Creator(s): It is not clear who actually created Tether

Primary Industries: Cryptocurrency

General Information:

"Tether is a token backed by actual assets, including USD and euros. One Tether equals one underlying unit of the currency backing it, e.g., the U.S. Dollar, and is backed

100% by actual assets in the Tether platform's reserve account. Being anchored or "tethered" to real world currency, Tether provides protection from the volatility of cryptocurrencies. Tether enables businesses—including exchanges, wallets, payment processors, financial services, and ATMs—to easily use fiat-backed tokens on blockchains."[72]

#10 Stellar

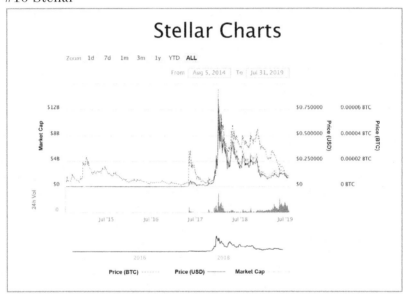

Currency Name: Stellar (XLM)
Year Created: 2015
Creator(s): Jeb McCaleb and Joyce Kim
Primary Industries: Payment System
General Information:

"Stellar is a platform that connects banks, payments systems, and people. Integrate to move money quickly, reliably, and at almost no cost."[73]

These graphs and additional information showcase the successful cryptocurrencies. While there are other successful cryptocurrencies, the CoinMarketCap top ten cryptocurrencies by market valuation provides examples for potential cryptocurrency investors. Similar to the stock market, the cryptocurrency market has "penny stocks" as well.

ARE CHEAP CRYPTOCURRENCIES TOO GOOD TO BE TRUE?

Although bitcoin and other cryptocurrencies are trading at high prices, there are many cryptocurrencies trading at very low prices.[74] There are investors looking for inexpensive cryptocurrencies in hopes they will dramatically increase in value. This approach comes with greater risks than the already volatile cryptocurrency market. Low priced cryptocurrencies, many under one tenth of one cent, have the potential to appreciate greatly in price; however, most will likely remain at the same low price or go out of business. Investing in cryptocurrencies without an accurate accounting for major investment factors such as track record, market capitalization, and circulating supply is extremely inadvisable.

SUMMARY

CoinMarketCap.com was the source of most of the information in this chapter; however, there are other online resources that provide useful information for cryptocurrency investors. It is advisable to use multiple resources to verify relevant information before investing in the cryptocurrency. Use CoinMarketCap.com, or other online cryptocurrency resources, to determine the legitimacy of the cryptocurrency. Once you have identified cryptocurrencies for possible investment, thoroughly research each cryptocurrency. Only then will you be able to make an educated investment.

SUCCESS STRATEGY ACTION ITEMS

1. CoinMarketCap.com contains a lot of links with a lot of information. Use the information to learn more about the individual cryptocurrencies.

2. The CoinMarketCap.com cryptocurrency graphs are an important feature that tells the trading history of a cryptocurrency. While graphs cannot help you determine the future performance of cryptocurrencies, they provide a useful historical perspective when deciding your investment strategy. When you are ready to invest, use the graphs as a source of information to formulate your investment plan and buy and sell decisions.

3. Visit the websites of the top ten cryptocurrencies to learn some of the best business practices in cryptocurrencies. While each site is very different in format and content, each site provides a basic understanding of the challenges each cryptocurrency was designed to solve. It may take more than one visit to fully understand, so repeat visits and continuous research is recommended.

4. Create other cryptocurrency wallets. Chapter 1 introduced the bitcoin wallet, but now CoinMarketCap.com serves as a central hub to visit cryptocurrency sites and create wallets for any active cryptocurrency. Use CoinMarketCap.com to access the cryptocurrency website where you will find a link or icon to create a specific cryptocurrency wallet.

[3]

CRYPTOCURRENCY INVESTMENT
INFORMATION

"Virtual Currencies may hold long-term promise, particularly if the innovations promote a faster, more secure, and more efficient payment system." [75]

—Ben Bernanke, Former Federal Reserve Chairman

ALTHOUGH MR. BERNANKE refers to cryptocurrencies as "virtual currencies," his quote points out the importance of payment systems in bitcoin and other cryptocurrencies. It also points out the three components of a payment system that will make it a success if all three are present: faster, secure, and efficient. If any one of these three components are lacking, or worse, missing, the currency will surely fail.

For example, let's analyze "faster." Currently, it takes at least ten minutes to finalize a bitcoin transaction. Other cryptocurrency innovators realized a ten-minute transaction time was too long for retailers and customers, so they designed blockchains with faster transaction times. For example, Litecoin has a minimum two-minute and thirty-second block-to-block transaction time.[76] In theory, reduced transaction times will increase the incentives for companies to accept bitcoin and other cryptocurrencies, and increase their market share on cryptocurrency exchanges. As such, this chapter focuses on cryptocurrencies as investments.

In theory, reduced transaction times will increase the incentives for companies to accept bitcoin and other cryptocurrencies, and increase their market share on cryptocurrency exchanges. As such, this chapter focuses on cryptocurrencies as investments.

CRYPTOCURRENCY INVESTMENT IS NOT FOR EVERYONE

As stated on the copyright page, the cryptocurrency investment analysis in this book is only a historical analysis and must not be used as the basis for any investment decisions. Nothing in this book should be construed as cryptocurrency investment or financial advice. Cryptocurrency investment is not for new, conservative, or faint of heart investors. Never invest in anything you do not understand. If you have not previously invested in stocks, currencies, or commodities, you should think twice about investing in cryptocurrencies. Cryptocurrencies are highly volatile investments!

In one conversation with a practicing physician, as we casually walked through a major U.S. city, he told me he invested a few thousand dollars in bitcoin in 2015. Then an interesting conversation followed.

"What was the bitcoin purchase price?" I asked him.

"About $50 per bitcoin," he replied.

He would not tell me how much he invested, but let's conservatively assume that "a few thousand" is $3,000, which was 60 bitcoin.

"I just pulled up CoinMarketCap.com, and right now, bitcoin is trading at $2,800 per coin. Congrats!" I then used my iPhone calculator to calculate that sixty bitcoin at $2,800 means $168,000 in profit. I asked him, "Do you realize you have $168,000 worth of bitcoin?"

"Yeah, it's too bad I sold them for a loss a few weeks after I bought them."

I literally stopped walking and looked at him. "What? Why did you sell?"

"I was constantly looking at the bitcoin prices online and could not take the large swings up and down, so I sold all my bitcoin."

The moral of this story is that cryptocurrency investing is not for everyone.

MY CRYPTOCURRENCY INVESTMENT STORY

After my initial cryptocurrency research provided enough information for me to decide to invest in cryptocurrencies, I had to learn how to select and purchase cryptocurrencies for investment and choose a cryptocurrency exchange.

Since I traded stocks online for over twenty years, I was very comfortable trading cryptocurrencies online. The next step was to understand the protocols for funding cryptocurrency exchanges. As discussed in Chapter 1, some cryptocurrency exchanges were hacked and were forced to close at a significant loss for their clients. I had to choose my cryptocurrency exchange carefully.

In 2016, most of the cryptocurrency exchanges were based in foreign countries, which required funding with local currencies rather than U.S. Dollars. This created an initial funding challenge, as I preferred to fund my cryptocurrency exchange account with U.S. Dollars rather than changing to another currency to fund my account.

At the time, Coinbase was the only cryptocurrency exchange that could be funded with U.S. Dollars and be linked directly to a U.S. bank account. But before I decided to choose Coinbase, I opened non-funded accounts on a number of cryptocurrency exchanges including Bitstamp, Bittrex, and Kraken.

Most cryptocurrency exchanges take your identity very seriously. You can review the specifics to open each account on each platform, but I will make three general comments about these platforms.

1. I was surprised at the extent of photographic information each platform required. Most exchanges requested a "selfie" with your government issued ID.

2. All of the cryptocurrency exchanges have at least a two-tier sign-in process. Tier one is the username and password. Tier two is the use of an authenticator app, most commonly Google Authenticator.

3. None of the cryptocurrency exchanges requested my social security number. This was a great comfort as hackers have stolen social security numbers from large organizations and government agencies.

CHOOSE THE CRYPTOCURRENCY EXCHANGE THAT WORKS FOR YOU

While Coinbase is very user friendly and has the capacity to link directly to a U.S.-based bank account, it also has limitations.

When I started trading on Coinbase, they only traded two cryptocurrencies: Bitcoin and Ethereum. As of the time of publication, Coinbase expanded the cryptocurrencies on their trading platform to include:

- Bitcoin (BTC),
- Bitcoin Cash (BCH),
- Ethereum (ETH),
- Ethereum classic (ETC),
- Litecoin (LTE),
- Ripple (XRP),
- Stellar lumens (XLM),
- Zcash (ZEC),
- Basic attention token (BAT),
- USD Coin (USDC),
- Augur (REP),
- Ox (ZRX), and
- Dai (DAI).[77]

Most of these cryptocurrencies are very well established and most have already experienced considerable increase in value. However, for traders interested in investing in "the next big" cryptocurrency, Coinbase provides limited options. By comparison, Binance trades 150 cryptocurrencies—far more than Coinbase.[78]

Ultimately, you have to choose the trading platform that best suits your investment strategy, but there are other ways to invest in bitcoin.

CRYPTOCURRENCY AUTOMATIC TELLER MACHINES (ATMS)

Cryptocurrency ATMs are another option to purchase or invest in bitcoin. Bitcoin ATMs were the first cryptocurrency ATMs, but now there are cryptocurrency ATMs for: Bitcoin Cash, Ether, Dash, and Litecoin. Cryptocurrency ATMs do not require a computer or cellphone and they accept fiat currencies. After the cryptocurrency purchase, the cryptocurrency ATM provides a receipt with important information about the cryptocurrency purchase. Afterwards, purchasers can transfer the cryptocurrency to another wallet for use as a currency.

CRYPTOCURRENCIES ARE HIGHLY VOLATILE

It is important to discuss the volatility of cryptocurrencies before the cryptocurrency investment discussion. Some cryptocurrencies can move over 10% in one day, but let's put cryptocurrency investments in perspective. According to Ycharts.com, the 2018 rate of return for the S&P Price Index was -4.38%.[79] That means that cryptocurrencies are highly volatile and can move the same percentage in a few days as the S&P moves in one year. For example, BitcoinExchangeGuide.com compiled a list of historical bitcoin corrections as shown in figure 6.[80]

In 2018, bitcoin hit a low of $3,179.69 USD, a 60.7% decline from the $8,094.80 USD and a total decline of 83.83% from the high of $19,666.00 USD Although this list of corrections focuses on bitcoin, all of the other cryptocurrencies experienced similar declines. For example, in 2018, Ethereum reached a high of $1389.18 USD and a low of $86.32 USD.[81] These rapid declines changed the way many small and institutional investors viewed bitcoin and other cryptocurrencies and the pulled their money out of the cryptocurrency market.

Correction Start Date	Correction End Date	# Days in Correction	Bitcoin High Price	Bitcoin Low Price	% Decline	$ Decline
2012	2012	16	$7.38	$3.80	-49%	-$3.58
2012	2012	3	$16.41	$7.10	-57%	-$9.31
2013	2013	2	$49.17	$33.00	-33%	-$16.17
2013	2013	3	$76.91	$50.09	-35%	-$26.82
2013	2013	3	$259.34	$45.00	-83%	-$214.34
2013	2013	1	$755.00	$378.00	-50%	-$377
2013	2015	411	$1,163.00	$152.40	-87%	-$1010.60
2017	2017	16	$1,350.00	$891.33	-34%	-$458.67
2017	2017	3	$2,760.10	$1,850.00	-33%	-$910.10
2017	2017	35	$2,980.00	$1,830.00	-39%	-$1,150.00
2017	2017	14	$4,979.90	$2,972.01	-40%	-$2,007.89
2017	2017	5	$7,888.00	$5,555.55	-30%	-$2,332.45
2017	2018	48	$19,666.00	$8,094.80	-59%	-$11,571.20

Figure 6 Historical bitcoin corrections.

CRYPTOCURRENCY BUBBLE AND CRASH?

On March 17, 2010, the now defunct BitcoinMarket.com started trading bitcoin at .003 cents per coin.[82] Bitcoin was not even worth one cent per coin. As of June 12, 2019, bitcoin was trading at $7,946.77 USD.[83] Using the formula – (Current price - Original price)/(Original price x 100) = Rate of return, here is the actual equation: ($7,946.77 - .003)/(.003 x 100) = 265,942.233% rate of return. Yes, 265,942.233%! Of course, only the first investors or miners received this exponential rate of return, but this rate of return begs these questions: Is this a cryptocurrency bubble? Will there be a cryptocurrency crash? And with decreases of 57%, 59%, 83%, and 87%, are these bitcoin crashes?

According to Investopia.com, a crash does not have a specific percentage market decrease, but it is most commonly associated with a double digit decrease in the market in a couple of days. By this definition, these are bitcoin crashes.

Other than the high rate of return of cryptocurrencies, one of the reasons the cryptocurrency bubble is prominent in the media is because of the dot-com bubble in 2000. While the term "bubble" is descriptive of the results when an investment rapidly increases and rapidly deflates, it is helpful to provide a financial definition to a "bubble." Investopia.com defines a bubble as:

> *An economic cycle characterized by rapid escalation of asset prices followed by a contraction. It is created by a surge in asset prices unwarranted by the fundamentals of the asset and driven by exuberant market behavior. When no more investors are willing to buy at the elevated price, a massive selloff occurs, causing the bubble to "deflate."[84]*

The Bitcoin Exchange chart clearly shows the rapid escalation in bitcoin prices followed by a massive sell off that burst the bitcoin bubble.

By comparison, in the dot-com bubble on March 10, 2000, the NASDAQ hit a peak of $5,048.62, but then after a massive investor sell off on October 10, 2002, the NASDAQ hit bottom at $1,139.90, an 76.81% loss.[85] In 2013 alone, bitcoin showed it's volatile nature as it crashed more than 80% twice. There are additional commonalities between the dot-com and cryptocurrency bubble, but there are also fundamental differences.

Commonalities:

1. The most important commonality is that both the dot-com companies and cryptocurrency companies are both based on new technology.

2. Dot-com stocks and cryptocurrencies very quickly attracted billions of dollars in investment capital resulting in a very rapid increase in price, but this commonality also has

a very important difference, which we will discuss in the differences section that immediately follows.

3. People invested in dot-com and cryptocurrencies because of fear of missing out (FOMO).

Differences:

1. The dot-com bubble was based on stocks listed on national stock exchanges and the cryptocurrency bubble was based in a decentralized, non-regulated environment on an international scale. This is much broader than investing in a stock.

2. The dot-com crash ripple effect impacted the major stock exchanges while the cryptocurrency crash was mostly confined to cryptocurrency exchanges. Cryptocurrencies are still years away from becoming a mainstream investment.

3. Traditional stock investing is based on fundamentals (revenues, earnings, profit and loss, and other indicators), whereas cryptocurrency investing does not have established fundamentals as of yet.

SUCCESSFUL COMPANIES AFTER THE DOT-COM CRASH

The dot-com story is usually focused on business failures, but there is another side to the story: Businesses that thrived since the dot-com crash. According to Investors.com, many companies recovered from the dot-com crash and the article focused on ten tech companies. Seven of the ten companies still trade under the same stock symbol. Three of the ten companies: ARM, Sandisk, and Priceline were either purchased by another company or renamed. The seven successful dot-com companies below are listed from lowest price during the peak of the dot-com mania to highest stock price.[86]

STOCK	LOW	HIGH
eBay (EBAY)	$4.57	$42.86
Adobe Systems (AD-BE)	$16.70	$284.25
Intuit (INTU)	$22.16	$261.41
Oracle (ORCL)	$15.54	$55.33
ASML (ASML)	$15.71	$214.00
Amazon (AMZN)	$42.70	$2,013.00
IBM	$81.60	$213.30

Figure 7 Historic dot-com stock comparison

This historic dot-com stock comparison is important to the cryptocurrency crash as the 2018 cryptocurrency crash does not mean the end of cryptocurrencies. Similar to the dot-com companies, cryptocurrency companies with a strong management team, valuable business solutions, and strong financials will likely survive the cryptocurrency crash. In fact, in 2019, many cryptocurrencies are on a path to recovery.

If cryptocurrency volatility is a concern, the next section introduces paper trading as a method to experience cryptocurrency investment without risking money.

PRACTICE TRADING BEFORE YOU ACTUALLY TRADE

If you want to practice trading cryptocurrencies before you invest any money, then paper trading is the perfect way to start. Paper trading is a practice where you select an investment (futures, stocks, commodities, etc.), determine the desired price, determine the amount of "paper" money to invest, and track your paper trading investment daily to determine your paper trading profit and loss. Your "paper balance" will increase and decrease as the cryptocurrency price increases and decreases. While you are not making or losing any actual money, tracking your paper trading profit and losses will allow you to become more comfortable with investing. I learned paper trading in the 1980s when I started to invest in stocks, and every time I invest in a new type of investment I always start with paper trading. Since I was a new cryptocurrency investor, I went back to my investment basics and started paper trading cryptocurrencies and keeping a paper-trading journal.

Cryptocurrency paper trading was a wild ride! All of the cryptocurrencies were moving up and down at an incredible pace. In December 2016, bitcoin was trading around $1,200. As a new investment, a $1,200 per bitcoin investment was out of my comfort zone, so I was content to watch bitcoin for a bit longer. Then paper trading showed its value. A few weeks after I learned about bitcoin, it dropped dramatically from $1,200 to $700; a 58% drop. Then the most amazing thing happened, within a few weeks, bitcoin recovered all of its losses and surpassed the previous high of $1,200. At that point I decided to stop paper trading and invest in cryptocurrencies. The final step of my paper trading was to review my cryptocurrency paper-trading journal to identify trends and trading patterns.

CRYPTOCURRENCY PAPER-TRADING JOURNAL

There are many ways to keep a Paper-Trading Journal: spreadsheet, audio recordings, Microsoft Word or Excel, screen shots of the investment, or online paper trading. As all methods are effective, select the method that best suits your learning and investment style. If your paper-trading needs are best met by an online paper-trading platform, there are apps that allow users to invest in dozens of cryptocurrencies by creating a play account, which is similar to paper trading. Coinseed.co is one of those apps.

For the purpose of this book, we will use a Microsoft Word table as an example.

The following paper-trading table sample tracks one week of trading for three cryptocurrencies.

PAPER-TRADING TABLE SAMPLE

NAME OF CRYPTOCURRENCY—BITCOIN (BTC)

NAME	SYMBOL	AMOUNT INVESTED	DAILY PRICE	VOLUME	DATE	TIME	% CHANGE
Bitcoin	BTC						
Bitcoin	BTC						
Bitcoin	BTC						
Bitcoin	BTC						
Bitcoin	BTC						
Bitcoin	BTC						
Bitcoin	BTC						
IMPORTANT NOTES:							

Figure 8 Bitcoin paper-trading table sample

NAME OF CRYPTOCURRENCY—ETHEREUM (ETH)

NAME	SYMBOL	AMOUNT INVESTED	DAILY PRICE	VOLUME	DATE	TIME	% CHANGE
Ethereum	ETH						
Ethereum	ETH						
Ethereum	ETH						
Ethereum	ETH						
Ethereum	ETH						
Ethereum	ETH						
Ethereum	ETH						
IMPORTANT NOTES:							

Figure 9 Ethereum paper-trading table sample

NAME OF CRYPTOCURRENCY—LITECOIN (LTE)

NAME	SYMBOL	AMOUNT INVESTED	DAILY PRICE	VOLUME	DATE	TIME	% CHANGE
Litecoin	LTE						
Litecoin	LTE						
Litecoin	LTE						
Litecoin	LTE						
Litecoin	LTE						
Litecoin	LTE						
Litecoin	LTE						
IMPORTANT NOTES:							

Figure 10 Litecoin paper-trading table sample

PAPER-TRADING KEY

- Name—The actual name of the cryptocurrency.
- Symbol—The trading symbol of the cryptocurrency.
- Amount Invested—The amount invested to each cryptocurrency.
- Daily Price—The actual price of the cryptocurrency as listed by the cryptocurrency trading platform.
- Volume—The number of coins bought and sold in a 24-hour period.
- Date—The date the cryptocurrency is tracked in the paper-trading tracking sheet.
- Time—The time the cryptocurrency is recorded in the paper-trading tracking sheet.
- % Change—The difference in price, calculated by percentage, from the current day against the prior day.
- Important Notes—Use this section to record information that helps in the decision to buy, sell, or continue to hold on to an

investment including: recent news, new regulations, new commercial adopters, etc.

BLANK PAPER-TRADING TABLE

NAME	SYMBOL	AMOUNT INVESTED	DAILY PRICE	VOLUME	DATE	TIME	% CHANGE
IMPORTANT NOTES:							

Figure 11 Blank paper-trading table

The above is a blank journal for your use (see figure 11). A paper-trading journal is more than a spreadsheet to track names, dates, and numbers; it is a method to create a disciplined trading routine. Similar to most activities in life, disciplined people experience higher levels of success for any activity. For example, a few potential cryptocurrency investors expressed interest in cryptocurrency investing. I advised them to start paper trading and provided the information they needed. Paper trading excited them and they decided to paper trade together. A few weeks later I ran into the same investors at another event.

"How is your paper trading going?" I asked the potential investors.

"We paper traded, but then we lost interest."

"So you decided not to invest in cryptocurrencies?"

"We would love to," they said, "but we just don't have the time to learn about cryptocurrencies."

"It's a discipline," I told them. "You have to set aside time to research and understand the topic."

"How did you stick to your paper trading program?"

"I set aside research and investing time everyday," I told them. "I do not want this opportunity to pass me by and regret it twenty years from now."

Since they did not have the discipline to paper trade, they decided not to invest. Paper trading showed them they did not have the desire to really trade cryptocurrencies, they were just fascinated by the success of other investors. This is still a good lesson to learn.

I shared with them that I had to discipline myself to paper trade every day and make it a part of my daily routine. They all considered my comments and decided to start paper trading in cryptocurrencies...again.

THE END OF PAPER TRADING AND THE START OF INVESTING

After paper trading, I started investing in cryptocurrencies by purchasing Ethereum and Litecoin. I liked both cryptocurrencies, both were far more affordable than bitcoin, and they were increasing in price at an amazing rate. Ultimately, I also invested in bitcoin for two reasons:

1. I wanted to diversify my cryptocurrency investment portfolio, and
2. Bitcoin is the most versatile cryptocurrency against which all other cryptocurrencies are measured and can be exchanged into bitcoin.

At the time, Coinbase only traded three cryptocurrencies: bitcoin, Ethereum, and Litecoin. Coinbase added Litecoin during my paper trading.

Most of the time, bitcoin, Ethereum, and Litecoin, the largest cryptocurrencies by market share, move up and down together. Occasionally, one of the three cryptocurrencies reports major progress in adoption of their technology or obtains a major commercial user, resulting in an independent increase in value. For example, Overstock.com was one of the first adopters of bitcoin and both companies' investment value increased when Overstock.com announced they were accepting bitcoin as an online method of payment.

Smart investors have a sound investment strategy for every investment, including cryptocurrencies, which will increase their chances of success. Investing in cryptocurrencies to see what happens is not a good approach. A disciplined trader who buys and sells based on an established entry and exit strategy is less likely to trade based on emotions. Instead, they will look at trading history and trends within the trading patterns to make decisions. In other words, it is best to enter/purchase the cryptocurrency at a predetermined price and exit/sell at a predetermined price for a profit. One of the biggest mistakes investors make, myself included, is to let emotions control their investment decisions.

Create an investment strategy, stick with it, and don't change it unless you have a great deal of new information that makes your initial investment strategy unlikely to succeed. The following sections will provide a brief introduction to three basic investment strategies: buy and hold, dollar cost averaging, and day trading.

BUY AND HOLD

Buy and hold is the first investment strategy in our discussion because I believe it generally improves chances for investment success. In the buy and hold strategy, investors purchase a financial instrument and hold onto it for a period of time, let's say a month or more. There are a couple of reasons why buy and hold provides a greater chance of success.

1. Generally, buy and hold investors are less likely to sell the investment during a downturn. If the investment rebounds, and returns to the original purchase price, the investor did not incur any losses.

2. Generally, buy and hold investors are less tempted to sell the investment quickly as they have a long-term mindset and are less likely to have a emotional reaction to price changes.

DOLLAR COST AVERAGING

Dollar cost averaging is buying a fixed dollar amount of an investment on a regular schedule. The investment focus is on accumulating assets on a regular basis instead of trying to "time the market." For example, a dollar cost averaging investor will buy $100 in "cryptocoin" every two weeks.

Week One: Cryptocoin is trading at $1.00 and the investor purchases 100 cryptocoins.

Week Two: Cryptocoin is trading at 50 cents and the investor purchases 200 cryptocoins.

Week Three: Cryptocoin is trading at $1.75 and the investor purchases 75 cryptocoins.

Dollar Cost Averaging Summary: Over three weeks, the dollar cost averaging investor spent $300 to purchase 375 cryptocoins.

Dollar cost averaging takes the emotion and fear out of investing because market direction in the short-term is less important when the investment strategy is to purchase a fixed dollar amount of cryptocurrencies over a period of time. If the cryptocurrency investment falls in value during a market downturn, the dollar cost averaging investor buys more coins at a lower price.

DAY TRADING

Day trading is exactly what it sounds like, buying an investment and selling it in the same trading day or in a very short period of time. Day traders are looking for a quick return on investment. Day trading is very risky with normal investments and even riskier for cryptocurrencies. Many long-term investors compare day trading to gambling, but many day traders have their own techniques to guide their investment decisions. Day traders also need to invest in trades with a high trading volume so they can sell the investment to another buyer to complete the trade. If the trading volume is low, they may have a problem selling their investment. Another factor day traders consider is the cost of placing the trade. If a trading platform charges $5 for each trade, the investment has to increase $5 for the trade to break even.

There are many more investment strategies and investment resources to educate investors on investment strategies. I encourage you to consider investment strategies that match your investment risk profile. A financial advisor is another resource for help in this area.

CONTACT YOUR FINANCIAL ADVISOR BEFORE YOU INVEST

Financial advisors can be an important investment resource. Many investors rely on the investment advice of their financial advisors, even when investing in cryptocurrencies. I have had many discussions with interested cryptocurrency investors like the one that follows.

"If you were interested in investing in cryptocurrencies, why didn't you invest?" I'd ask.

"My financial advisor told me cryptocurrencies are a new technology and I should wait until it becomes more established."

Then I'd ask, "How much money would your financial advisor make if you invested in cryptocurrencies?"

(They would all give the same answer.) "They wouldn't make any money."

That's when they realized one of the reasons financial advisors probably did not encourage cryptocurrency investment: the financial advisor would not make any commission off the transaction.

For the record, *Blockchain or Die* recommends seeking the advice of a financial advisor, but the financial advisor's advice should not serve as the *only* decision point. After you talk to a financial advisor, conduct thorough research, and come to your own conclusions about cryptocurrency investment.

SUMMARY

While I was traveling between blockchain conferences, I had the pleasure of sitting next to an older gentleman on the plane. After we ordered our drinks, I took out my draft of *Blockchain or Die* to start writing, and when he saw what I was doing, he struck up a conversation. This shortened version of the conversation on cryptocurrency investing is a good way to end this chapter.

"What are you working on?" the gentleman asked.

"I am writing a book on cryptocurrencies and blockchain technology."

He looked impressed. "Really? What do you do?"

"I am an attorney and a certified blockchain trainer," I replied.

"Is this bitcoin? I don't understand all of this bitcoin stuff." I provided a brief description of cryptocurrencies and blockchains. Then he said, "I don't like bitcoin because I can't hold it in my hand. I buy gold and silver because I can actually hold it in my hand." He literally smacked his hand so hard my hands felt it. "If I can't hold it, I won't buy it."

"That is a common sentiment," I told him. "Tell me, do you buy stocks?"

"Yes." He went on to list a number of stocks.

"How long have you been buying stocks?"

"Over fifty years," he said.

"So, in the 1960s, were you issued stock certificates?"

"Of course! I kept them in my safe."

"Do you buy stocks online now?"

"Of course!"

"Can you show me your stock certificates?" I painfully smacked my hand as loud as I could.

The gentleman laughed loudly. "You got me there! Good one."

Then I realized other people around us were listening and started to laugh as well. So I had to bring the learning point home.

"Finance and technology are always evolving," I began. "We can't get used to the way things used to be, otherwise we will never move to the future. Just like you switched from buying stocks through a broker to buying stocks online, without a stock certificate, if you want to think about investing in the future, you should consider buying investments that you can't hold in your hand, but rather can be held in your computer, phone, hard drive, or even a piece of paper. That is why the title of my book is *Blockchain or Die.*"

"A piece of paper like a stock certificate?" the gentleman asked.

"Actually, yes. Just never lose it!"

The gentleman literally presented a toast with his drink and ended by saying, "I am definitely going to buy your book when it comes out."

This is a good way to summarize cryptocurrency investing; it's the investing of the future and it is investing in your future. The most profitable and successful investments are ones that look to the future, not remain in the past.

SUCCESS STRATEGY ACTION ITEMS

1. Research cryptocurrencies to choose one that meets your investment needs. This includes:
   a. Looking at the cryptocurrency on CoinMarketCap.com.
   b. Visiting the website and social media pages of the cryptocurrency as well as reading the white paper.
   c. Reading blogs about the cryptocurrency.
   d. Researching other cryptocurrencies in the same industry.
   e. Attending conferences where the president or a senior officer is discussing the status of their cryptocurrency.
2. Try paper trading using the materials provided in this chapter. Paper trading is a discipline, so for the best re-

sults, take ten minutes out of every day to continue paper trading. Use the paper-trading tool in this chapter to track your cryptocurrency paper trades. Create your own paper-trading journal or find a paper-trading tool online. Paper trade until you are ready to invest in cryptocurrencies. If paper trading shows you are not comfortable with or ready to invest in cryptocurrencies—great—at least you did not lose any money in the process.

3. Open one or more cryptocurrency exchange accounts. You do not have to fund your account until you are ready. Take your time and learn how the trading account works and get use to reading the graphs and charts. Use the exchange account that best suits your trading strategy until you feel comfortable funding it (if you open more than one account).

4. Deposit fiat currency into the exchange account.

5. Determine your cryptocurrency investment strategy. There are many more investment strategies than the ones covered in this chapter. Research the investment strategies that fit your risk tolerances.

6. Purchase the cryptocurrency at the desired price. If price does not matter, you can purchase the cryptocurrency at market price.

7. Determine if you want to keep your cryptocurrency on the exchange account or transfer it to one of the wallets covered in the previous chapter.

8. Track the progress of your cryptocurrency. This one is critical. Don't just invest and forget to look at your investment. Similar to the paper trading exercises be sure to keep track of your investments.

9. Use CoinATMRadar.com to find a cryptocurrency ATM near your location.

10. Discuss your cryptocurrency investment plan with your financial advisor.

11. Finally, never invest in anything you do not understand.

The information provided in this chapter and the above step-by-step process will assist in your cryptocurrency investment experience.

[4]

THE REGULATION OF
CRYPTOCURRENCIES

The SEC is right to insist that the digital coins, such as bitcoin, which are replacement for sovereign currencies, such as the dollar, sterling, yen, and euro, are not securities. I believe the SEC is also right that tokens that act as digital assets are securities.[87]

—*Jay Clayton, SEC Chairman*

AS THIS CHAPTER SPECIFICALLY addresses the regulation of cryptocurrencies, it is appropriate to start the chapter with a quote from the Chairman of the Securities and Exchange Commission (SEC), the U.S. government agency that protects investors.

Chairman Clayton's quote addresses the past, present, and future of cryptocurrencies. The reference to "removes some of the uncertainty" summarizes the SEC's past position on cryptocurrencies in that companies improperly created or marketed initial coin offerings in violation of SEC regulations. The reference to "strengthens the overall proposition on cryptocurrencies" arguably shows the SEC's present day position, which is to support cryptocurrency investments in compliance with SEC regulations. The reference to "recognizing the scale and potential of bitcoin and other cryptocurrencies" provides possible insight into the SEC's position on the future state of cryptocurrencies. The SEC will play a critical role in the future of cryptocurrencies in the U.S. as SEC regulations and public statements will probably impact cryptocurrency prices.

WHY IS THE REGULATION OF CRYPTOCURRENCIES IMPORTANT?

Every investor, or potential investor, should have a basic understanding of the regulations governing their investment. This book exposes the reader to the basics of cryptocurrencies and blockchains and this chapter introduces the regulatory framework. Whenever I speak about cryptocurrencies, I always state: "never invest in anything you do not understand." In order to understand cryptocurrencies, it is important to understand the applicable regulations. Since regulators are in the early phases of regulating cryptocurrencies, there are frequent regulatory updates as well as court cases. Interested investors are strongly advised to use the resources provided in this chapter and the Appendix to stay abreast of the regulatory updates. Once again, let's start with bitcoin.

HOW DOES THE U.S. GOVERNMENT DEFINE BITCOIN?

The first, and much debated, regulatory question about bitcoin is "which U.S. federal agency has the jurisdiction to regulate bitcoin?" Bitcoin has been referred to as:

1. A commodity
2. An investment

3. A currency

4. Property

Four different U.S. Government Agencies separately regulate these areas.

1. A commodity—The Commodities and Futures Trading Commission[88]

2. A currency, but if it is a token, it may also be considered a security—The Securities and Exchange Commission[89]

> *Token—A digital unit designed with utility in mind, providing access and use of a larger crypto-economic system.*

3. A currency—Department of Treasury, Financial Crimes Enforcement Network,[90] and the U.S. Office of Foreign Assets Controls[91]

4. Property—The Internal Revenue Service[92]

The regulatory classification of bitcoin, as well as other cryptocurrencies, is critically important to the future of cryptocurrencies. For example, if bitcoin were classified as a "security," the SEC would require cryptocurrency exchanges, including foreign exchanges that sell to U.S. citizens, to register as an exchange. The SEC registration process would reduce the amount of exchanges for investors to buy and sell, which would likely lower the price. This is one example how the regulatory classification can impact cryptocurrencies and exchanges.

Another example includes Ethereum, which the SEC recently stated that Ether, the currency for Ethereum, is not a security. Ethereum is very different from bitcoin in many ways: bitcoin's primary design is to serve as a currency and Ethereum's primary design is to serve as an application platform.

The remainder of this chapter will address the regulatory considerations, including agency statements, investigations, and rulings. At the time of this writing, the U.S. Government has not designed a comprehensive cryptocurrency regulatory framework. As such, the following information is set forth by individual agencies as objectively as possible and supported by current versions of laws, regulations, and agency rulings. After reviewing this chapter, and conducting additional research, readers are encouraged to follow the status of the U.S. regulations of cryptocurrencies.

THE COMMODITIES AND FUTURES TRADING COMMISSION (CFTC) REGULATION OF CRYPTOCURRENCIES

As of the time of this writing, bitcoin is primarily considered a commodity. When people think of commodities, they usually think of:

- Metals—Gold, silver, or platinum
- Agricultural—Soybeans, wheat, or corn
- Energy—Heating oil, crude oil, or natural gas
- Livestock—Pork bellies, live cattle, or feeder cattle

Investopia.com defines commodity as "a basic good used in commerce that is interchangeable with other commodities of the same type."[93] All of the above commodities fit that definition, and so does bitcoin...somewhat.

> *Mining—The process of using computer power to solve a complex math problem presented by the crypto system, review and verify information, and create a new recording to be added to the blockchain.*

There are similarities. Commodities, such as those listed above, do not belong to one person or company. Anyone can: mine gold, grow soybeans, mine for oil, or grow cattle just like anyone can mine bitcoin. As explained in Chapter 1, bitcoin is open source and decentralized making it accessible to anyone who wants to mine bitcoin.

Bitcoin is clearly used in commerce as bitcoin can be used to purchase goods and services all over the world. Bitcoin is interchangeable as it is traded on cryptocurrency platforms. So from that basic definition, bitcoin is a commodity.

Commodities are covered by the Commodity and Exchange Commission Act (CEA), which is enforced by the CFTC. The website at CFTC.gov states:

> "The mission of the Commodities and Futures Trading Commission (CFTC) is to foster open, transparent, competitive, and financially sound markets. By working to avoid systemic risk, the Commission aims to protect market users and their funds, consumers, and the public from fraud, manipulation, and abusive practices related to derivatives and other products that are subject to the *Commodity Exchange Act (CEA)*."[94]

In December 10, 2014, testimony before the U.S. Senate Committee on Agricultural, Nutrition, and Forestry, Timothy Massad, Chairman of the Commodities Futures Trading Commission stated bitcoin is a commodity:

> "The CEA defines the term commodity very broadly so that in addition to traditional agricultural commodities, metals, and energy, the CFTC has oversight of derivatives contracts related to Treasury securities, interest rate indices, stock market indices, currencies, electricity, and heating degree days, to name just a few underlying products...Derivative contracts based on a virtual currency represent one area within our responsibility."[95]

The CEA very broadly defines commodity as:

"The term 'commodity' means wheat, cotton, rice, corn, oats, barley, rye, flaxseed, grain...motion picture box office receipts (or any index, measure, value, or data related to such receipts), and all services, rights, and interests (except motion picture box office receipts, or any index, measure, value or data related to such receipts) in which contracts for future delivery are presently or in the future dealt in."[96]

The most of the CEA covers agricultural and livestock considerations, but bitcoin is covered in the final section: "all services, rights, and interests ... in which contracts for future delivery are presently or in the future dealt in."[97] When the CFTC sued My Big Coin Pay, Inc. in the Eastern District of New York for fraud and false claims, Judge Weinstein denied My Big Coin's motion to dismiss ruling the CFTC has the power to prosecute fraud in virtual currency markets. The District Court used the same "any and all services" clause of the CEA to support their interpretation that My Big Coin was a commodity.[98] As discussed in the next section, the Securities and Exchange Commission regulates bitcoin-based investments.

SECURITIES AND EXCHANGE COMMISSION (SEC) REGULATION OF CRYPTOCURRENCIES

According to SEC.gov, "[t]he mission of the SEC is to protect investors; maintain fair, orderly, and efficient markets; and facilitate capital formation. The SEC strives to promote a market environment that is worthy of the public's trust."[99] It is the responsibility of the SEC to:

- Interpret and enforce federal securities laws;
- Issue new rules and amend existing rules;

- Oversee the inspection of securities firms, brokers, investment advisers, and ratings agencies;
- Oversee private regulatory organizations in the securities, accounting, and auditing fields; and
- Coordinate U.S. securities regulation with federal, state, and foreign authorities.[100]

The most common laws the SEC uses for enforcement over the securities industry include the Securities Act of 1933[101] and the Securities Exchange Act of 1934.[102]

None of the above laws specifically use the words "bitcoin" or "cryptocurrencies," but the SEC has ruled they cover most investments, including cryptocurrencies. According to CNBC.com, the SEC Chairman clearly stated the SEC is not going to change the definition of a security: "We are not going to do any violence to the traditional definition of a security that has worked for a long time."[103] This statement stabilizes the legal and regulatory standing of cryptocurrencies, but the SEC's regulation on cryptocurrencies goes far beyond a legal definition.

The "Howey Test" is one of the tools the SEC uses to determine if a transaction is an investment. The Howey Test is based on the Supreme Court case SEC v. J.W. Howey Co., where the Howey Co. sold parts of their citrus land to raise capital to finance additional development. The SEC filed an injunction since Howey Co. did not register the land transactions. The Supreme Court found Howey Co. violated securities laws because the purchasers of the land were spectators that did not possess the "knowledge, skills, and equipment necessary for the cultivation of citrus trees." As such, the purchasers were investing in Howey Co.'s expertise to work on the citrus land, which would increase the value of their investment.[104] The Howey Test can apply to any industry and any type of investment.

The SEC uses three steps to apply the Howey test:

1. The Investment of Money—The digital asset is acquired in exchange for value.
2. Common Enterprise—A business or investment that is managed by third parties.
3. Reasonable Expectation of Profits Derived from the Efforts of Others—This is exactly what it sounds like, participants have an expectation of return on investment.[105]

How does the Howey test relate to cryptocurrencies? The answer is: "it depends on the cryptocurrency investment offering." Similar to Howey, if the cryptocurrency investment offering is a speculative investment and the investor is relying on the expertise of the management team to increase the value of their investment, the SEC will likely find the cryptocurrency is an investment and subject to SEC regulations.

The SEC is already applying the Howey Test in their efforts to investigate and enforce Initial Coin Offerings (ICOs). The SEC has already investigated a number of cryptocurrency investment offerings; one of the first investigations was the decentralized autonomous organization (DAO).

THE SEC DECENTRALIZED AUTONOMOUS ORGANIZATION (DAO) INVESTIGATION

The SEC's Report of Investigation found that crypto tokens offered and sold by a "virtual" organization known as "The DAO" were securities and subject to federal securities laws. The SEC Investigative Report confirms issuers of distributed ledger or blockchain technology-based securities must register offers and sales of such securities unless a valid exemption applies. The purpose of the registration provisions of the federal securities laws is to protect investors and ensure investment offerings include all the proper disclosures. Those participating in unregistered offerings also may be liable for violations of the securities laws.

This SEC Investigative Report is important because it is an official SEC statement that DAO tokens are securities. The SEC found offers and sales

> *Decentralized Autonomous Organizations (DAO)—A leaderless organization supported by a network of computers.*

of digital assets, ICOs, or "Token Sales" by "virtual" organizations are subject to the requirements of the federal securities laws. In coming to this conclusion, the SEC considered more than terminology, they also considered facts and circumstances of the offers and sales of digital assets.[106]

The Investigative Report confirms issuers of distributed ledger or blockchain technology-based securities must register offers and sales of such securities unless a valid exemption applies. Issuers participating in unregistered offerings also may be liable for violations of the securities

> *Initial Coin Offering (ICO)—When a new cryptocurrency or token generally becomes available for public investment.*

laws. Additionally, cryptocurrency exchanges that provide for trading in these securities must register unless they are exempt. The registration provisions of the federal securities laws ensures investors are sold investments that include all the proper disclosures and are subject to regulatory scrutiny for investors' protection.[107]

At the time this book is published, Congress has not passed any laws specifically related to cryptocurrencies. However, as shown in the following sections, the SEC has been overseeing and regulating initial coin offering (ICO) process.

SEC ACTIVELY INVESTIGATES INITIAL COIN OFFERINGS

ICOs are a method of fundraising that companies use to raise funds for their businesses. ICOs are similar to initial public offerings (IPOs) where companies sell stocks to raise capital. As the sale of stocks and IPOs fall under the jurisdiction of the SEC, and ICOs have been raising massive amounts of money, the SEC has been closely monitoring ICO activity.

The SEC is protecting investors and removing uncertainty in ICO sales by investigating potentially fraudulent ICOs. The SEC issued numerous investigative reports cautioning investors offering digital assets by virtual organizations are subject to the requirements of the federal securities laws. Such offers and sales, conducted by organizations using distributed ledger or blockchain technology, have been referred to as ICOs or "Token Sales." Whether a particular investment transaction involves the offer or sale of a security—regardless of the terminology or technology used—depends on the facts, circumstances, and economic realities of the transaction.

According to ICOData.io, companies started to use ICOs to raise funds in 2014. As shown in the following chart in figure 12, until 2018, ICOs exponentially increased in number and amount of funds raised.[108] [109] [110] [111] [112] [113]

Year	Number of ICOs	Total Amount Raised
2014	2	$16,032,802
2015	3	$6,084,000
2016	29	$90,250,273
2017	875	$6,226,689,449
2018	1257	$7,850,727,041
To May 2019	TBD	$231,039,000
Total	2,116	$14,189,783,565

Figure 12 Number of ICOs and funds raised 2014–2018

The total amount of funds raised from 2014 to 2019 is staggering; and created a new form of venture capital. The average amount raised per ICO is $67,059,446.86. But companies using ICOs as a funding source may be slowing down: the 2019 ICO data shows that as of May 2019, ICOs raised only $231,039,000. This is tracking to a lower amount of ICO funding for 2019. As of the time of publication, ICOdata.io did not specify the actual number of ICOs in 2019.

There are a number of ICO resources online including https://topicolist.com/ and https://www.coindesk.com/6-3-billion-2018-ico-funding-already-outpaced-2017/.

ICOs BACKED BY REAL ESTATE AND DIAMONDS?

In September 2017, the SEC charged a businessman and two companies with defrauding investors in a pair of so-called "ICOs" purportedly backed by investments in real estate and diamonds.

The SEC alleged the owners of REcoin were selling unregistered securities and non-existent digital tokens. According to the SEC's complaint, investors in REcoin Group Foundation and DRC World (Diamond Reserve Club) were told to expect sizeable returns from the companies' operations when neither had any real operations.

The owners marketed REcoin as "The First Ever Cryptocurrency Backed by Real Estate." REcoin allegedly mislead investors to believe they had a "team of lawyers, professionals, brokers, and accountants" who would invest REcoin's ICO proceeds into real estate when no experts had been hired or even consulted. REcoin owners also allegedly misrepresented they raised between $2 million and $4 million from investors when the actual amount was approximately $300,000.

According to the SEC's complaint, REcoin carried the scheme over to DRC World, which allegedly invested in diamonds and obtained discounts with product retailers for individuals who purchase "memberships" in the company. Despite their representations to investors, the SEC alleged REcoin and DRC World did not purchase any diamonds or engage in any business operations, but continued to solicit investors and raise funds as though they had purchased diamonds.[114]

FALSE PROFIT CLAIMS

In December 2017, the SEC obtained an emergency asset freeze to halt a fast-moving ICO fraud that raised up to $15 million from thousands of investors by falsely promising a thirteenfold profit in less than a month. The SEC filed charges against PlexiCorps alleging they marketed and sold securities called PlexiCoin on the internet to investors claiming PlexiCoin would yield a 1,354 percent profit in less than twenty-nine days.[115]

INVESTING IN A DECENTRALIZED BANK

In January 2018, the SEC obtained a court order halting an allegedly fraudulent ICO that targeted retail investors to invest in the world's first "decentralized bank." According to the SEC's complaint, AriseBank used social media, a celebrity endorsement, and other marketing tactics to raise what it claims to be $600 million of its $1 billion goal in just two months.

AriseBank co-founders allegedly offered and sold unregistered investments in their "AriseCoin" cryptocurrency by offering a variety of consumer-facing banking products and services using more than 700 different virtual currencies. AriseBank claimed it developed an algorithmic trading application that automatically trades in various cryptocurrencies.

AriseBank allegedly falsely stated it purchased an FDIC-insured bank, which enabled it to offer FDIC-insured accounts and an AriseBank-branded Visa card to customers to spend at any of the 700-plus cryptocurrencies.[116]

TOKENS AS UNREGISTERED SECURITIES—THE SEC AND MUNCHEE, INC.

In December 2017, Munchee, Inc. sold digital tokens to investors to raise capital for its blockchain-based food review service. Munchee, Inc. was seeking $15 million in capital to improve an existing iPhone app centered on restaurant meal reviews and create an "ecosystem" in which Munchee, Inc. and others would buy and sell goods and services using the tokens.

Munchee, Inc.'s website, and white paper alleged it would use the proceeds to create the ecosystem and eventually pay users in tokens for: 1) writing food reviews, and 2) selling both advertising to restaurants and "in-app" purchases to app users.

Munchee, Inc. emphasized investors would receive an increase in value of their tokens and create a secondary market for the tokens. Because of these, and other company activities, investors would have had a reasonable belief their investment in tokens could generate a return on their investment. SEC halted Munchee, Inc.'s ICO when they found the ICO was an unregistered security. As the SEC stated in the DAO Report of Investigation, a token can be a security based on the long-standing facts and circumstances test that includes assessing whether investors' profits are to be derived from the managerial and entrepreneurial efforts of others. In Munchee, the creation of a blockchain-based food service and "ecosystem" coupled with the emphasis that investors would increase in value qualified Munchee, Inc.'s ICO as an "unregistered security." According to the SEC's order, before any tokens were delivered to investors, Munchee, Inc. refunded investor proceeds.[117]

FALSE ADVERTISEMENT HAS CONSEQUENCES

The SEC entered an order finding that Crypto Asset Management LP (CAM) offered a fund that operated as an unregistered investment company by engaging in an unregistered non-exempt public offering and investing more than forty percent of the fund's assets in digital asset securities. CAM falsely marketed the fund as the "first regulated crypto asset fund in the United States." According to the SEC's order, CAM, raised more than

$3.6 million over a four-month period in late 2017 while falsely claiming the fund was regulated and filed a registration statement with the SEC. After being contacted by the SEC, CAM ceased its public offering and offered buy backs to affected investors.[118]

UNREGISTERED SECURITY-BASED SWAPS

The SEC filed charges against an international securities dealer and its Austria-based CEO for allegedly violating the federal securities laws in connection with security-based swaps funded with bitcoin.

According to the SEC's complaint, 1pool Ltd. a/k/a 1Broker, registered in the Republic of the Marshall Islands, and its CEO Patrick Brunner solicited investors from the United States and around the world to buy and sell security-based swaps. Investors could open accounts by simply providing an email address and a user name—no additional information was required—and could only fund their account using bitcoin. The SEC alleges an undercover Special Agent with the Federal Bureau of Investigation successfully purchased several security-based swaps on 1Broker's platform from the U.S. despite not meeting the federal securities laws discretionary investment thresholds. The SEC also alleges Brunner and 1Broker failed to transact its security-based swaps on a registered national exchange, and failed to properly register as a security-based swaps dealer.[119]

As a special note, the 1pool Ltd. SEC order shows the SEC is paying attention to international companies transacting business with U.S. citizens and is working with the FBI to expose activities that are not in compliance with SEC regulations.

Unregistered Broker-Dealers

The SEC found TokenLot LLC, a self-described "ICO Superstore," acted as unregistered broker-dealers. According to the SEC's order, TokenLot used its website to purchase digital tokens during initial coin offerings (ICOs) and engaged in secondary trading. TokenLot received orders from more than 6,100 retail investors and handled more than 200 different digital tokens, which the SEC found included securities. TokenLot's profits included trading profits and a percentage of the money raised for ICOs. TokenLot's activities needed registration with the SEC as broker-dealers, but they were not. In response to the SEC's investigation, TokenLot voluntarily began winding down and refunding investors' payments for unfilled orders.[120]

ICO Investors: Buyer Beware

The above SEC investigations show there are companies that will attempt to defraud investors. Before investing in an ICO, it is important to conduct detailed research, including a detailed analysis of the ICO white paper. As discussed below, companies fraudulently created ICOs to defraud investors or did not properly communicate the investment risks. These companies sold their ICO and promised a return on investment, but the SEC found a number of ICOs violated securities laws and some were fraudulent. It's the classic case of "buyer beware."

The Bitcoin Savings and Trust Ponzi Scheme

Uninformed bitcoin critics have referred to bitcoin as a "Ponzi scheme," but they are incorrectly using the term. A Ponzi scheme, named after the infamous swindler Charles Ponzi, is an investment scam based on the payment of promised returns to existing investors from funds contributed by new investors.

Ponzi scheme organizers attract new investors by promising to invest funds in opportunities that will generate high returns with little or no risk to the investor. Rather than investing in a legitimate investment opportunity, Ponzi scheme organizers attract new money to make promised payments to existing investors while keeping the "invested funds" for personal use. In a Ponzi scheme, the last investors lose their money because the Ponzi scheme is no longer attracting new investors to "invest new funds."

The following dialogue represents a common conversation I have had with people who believed bitcoin was a Ponzi scheme:

"I would never invest in bitcoin; it's basically a Ponzi scheme," one bitcoin critic said.

"Ponzi schemes are centralized and require the organizer or existing members to bring in new members so they can cash out," I said. "The magic of bitcoin is that it is decentralized and anyone can invest or sell their investment whenever they want."

"Bitcoin has everyone fooled—it is not the typical Ponzi scheme, but it is a new version of a Ponzi scheme."

"Tell me specifically how bitcoin is a Ponzi scheme."

"I just know that it is," the critic would say.

"Have a nice day."

The bottom line is when bitcoin critics give you that kind of response, you probably won't change their minds, and there are much better things you could be doing with your time, like researching cryptocurrency updates.

The SEC investigates and prosecutes many Ponzi scheme cases each year to prevent new victims from being harmed and to maximize recovery of assets to investors.

As with many frauds, Ponzi scheme organizers often use the latest innovation, technology, and product or growth industry to entice investors and give their scheme the promise of high returns. But this is not always the case. Bernie Madoff, the orchestrator behind one of the greatest Ponzi schemes in history, used blue chip stocks as the investment vehicle for his investors. Madoff investors were less skeptical of established and stable blue chip investments. Regardless of the investment, it's only a matter of time before the Ponzi scheme fails.[121]

Based on this definition, bitcoin as an organization and a currency is not a Ponzi scheme, but Bitcoin Savings and Trust was a Ponzi scheme. Bitcoin Savings and Trust offered a high rate of return, 7% per week, and new customer deposits were used to pay profits to current customers. Using new customer's deposits to pay Ponzi scheme profits to current customers is a Ponzi scheme in its purest form. In August 2012, the U.S. Government shut down Bitcoin Savings and Trust and indicted the organizer. The government found the Bitcoin Savings and Trust scheme cost victims more than 265,678 bitcoin.[122]

WHAT ARE "AIRDROPS"?

Historically, airdrops date back to World War II when the military dropped supplies from airplanes to troops when they could not land planes. In the blockchain world, "airdrops" are where blockchain companies "drop" a few free crypto tokens to parties who perform marketing tasks for the tokens, including liking, following, and posting on social media. Once the marketing tasks are performed, the crypto tokens are deposited into the airdrop user's wallet to complete the airdrop transaction.

Airdrops are free and worth very little (if anything) when they are dropped, so most cryptocurrency users do not consider them investments. But in the SEC memo, "Framework for "Investment Contract Analysis for Digital Assets," the SEC found that airdrops were investments because they may constitute a sale or distribution of securities when it is distributed to holders of another digital asset, typically to promote its circulation.[123]

Given these and other scenarios, be sure to determine if the cryptocurrency has registered with the SEC before giving the company any of your information.

UTILITY TOKENS

As stated earlier in the chapter, the use of ICO to raise capital may be decreasing. Utility tokens are another way businesses raise capital. Businesses create utility tokens to power and finance their blockchain. Businesses hold back utility tokens and release the rest to finance their business in hopes the tokens will increase in value. By definition, utility tokens are not securities as they have a role, features, and purpose in the organization. As they are not securities, the SEC does not enforce them.[124]

SEC SUMMARY

The SEC has a difficult balancing act to protect U.S. citizens from businesses seeking to illegally take investor's money while regulating a new industry; however, the cryptocurrency community is also considering the same balance. Some cryptocurrency investors are against cryptocurrency regulations while others use the examples of inadequate disclosures and investor fraud to support the need for regulations. The SEC will play a major role in determining the success of cryptocurrencies and ICOs. Thus far, the SECs role has been mostly protective and reactive, but if Congress passes cryptocurrency laws for SEC enforcement or the

SEC issues cryptocurrency regulations, the SEC will have a greater proactive impact on cryptocurrencies and ICOs.

Congress has started proposing cryptocurrency legislation from a tax perspective. The first major piece of legislation is the Token Taxonomy Act to be covered in the IRS section below.

CRYPTOCURRENCIES AND THE INTERNAL REVENUE SERVICE

The Internal Revenue Service (IRS) is another U.S. government agency with regulatory authority over one aspect of cryptocurrency investment: cryptocurrency profit and losses. The IRS views cryptocurrencies, which the IRS calls "virtual currencies" as "property" and ruled on the taxation of cryptocurrencies by stating: "The sale or other exchange of virtual currencies, or the use of virtual currencies to pay for goods or services, or holding virtual currencies as an investment, generally has tax consequences that could result in tax liability." According to the IRS, virtual currency is a digital representation of value that functions as a medium of exchange, a unit of account, and/or a store of value.[125]

IRS Notice 2014-21 is one of the most important IRS documents related to cryptocurrencies and the frequently asked questions (FAQs) are an excellent resource to answer cryptocurrency tax liability questions.[126] For example, two of the FAQs listed on the IRS website (https://www.irs.gov/pub/irs-drop/n-14-21.pdf) are quoted below.[127]

> Q-1: How is virtual currency treated for federal tax purposes?

> A-1: For federal tax purposes, virtual currency is treated as property. General tax principles applicable to property transactions apply to transactions using virtual currency.

Q-3: Must a taxpayer who receives virtual currency as payment for goods or services include in computing gross income the fair market value of the virtual currency?

A-3: Yes. A taxpayer who receives virtual currency as payment for goods or services must, in computing gross income, include the fair market value of the virtual currency, measured in U.S. dollars, as of the date that the virtual currency was received. See Publication 525, Taxable and Nontaxable Income, for more information on miscellaneous income from exchanges involving property or services.

The rest of the FAQs cover other cryptocurrency tax issues including:

- Determining fair market value of virtual currencies
- Determining gain or loss upon exchange of a virtual currency
- Tax implications of successful mining

IRS notice 2014-21 is a must read for cryptocurrency investors, but this notice is the first of many notices the IRS will issue as cryptocurrencies mature and create new revenue models.

UNITED STATES CONGRESS' LETTER TO THE INTERNAL REVENUE SERVICE

The path to clear IRS guidance for taxing cryptocurrencies is ongoing. A coalition of U.S. Congressional Representatives in the Congressional Blockchain Caucus sent an open letter to the David Kautter, IRS Commissioner, pointing out the IRS has been cracking down on businesses and individuals to pay taxes for use of digital assets without issuing sufficient guidance or clarification.

The Congressional Blockchain Caucus specifically requested "additional guidance on the tax consequences and basic reporting requirements for transactions involving virtual currencies."[128] The Congressional Blockchain Caucus recommended the IRS take action to ensure increased taxpayer compliance with Notice 2014-21 and publish provisional guidelines on how digital transactions and investments should be handled when U.S. citizens file their taxes.

As the use of virtual currencies as a currency or an investment increases, the IRS will have to ensure clarity in the IRS regulations and communications so it does not limit the use of cryptocurrencies. These proposed provisional guidelines should provide taxpayers with a greater understanding of their tax obligations in the use of cryptocurrencies.

THE TOKEN TAXONOMY ACT

The Token Taxonomy Act, a bipartisan bill, seeks to provide legislative and regulatory clarity for a fundamental question in the cryptocurrency world: Are cryptocurrencies securities?[129] The Taxonomy Act defines a "digital token" in the context of a number of existing securities laws, most importantly, the Securities Act of 1933 and the Securities Exchange Act of 1934.[130] As stated earlier in the chapter, the SEC Chairperson has publicly stated he does not intend to change the existing SEC laws, so this bill may be part of long-term strategy to place proposed definitions on the record for future SEC Chairpersons, or it may be persuasive reference for the SEC during investigations. Either way, the Token Taxonomy Act is a step in the right direction to clarify the definition of a cryptocurrency.

REGULATION OF CRYPTOCURRENCY RELATED FINANCIAL CRIMES

The Department of Treasury Financial Crimes Enforcement Network's (FINCEN) mission is to "safeguard the financial system from illicit use, combat money laundering, and promote national security through the strategic use of financial authorities and the collection, analysis, and dissemination of financial intelligence."

The U.S. Government uses FINCEN to regulate Know Your Customer (KYC) and Anti-Money Laundering (AML) matters. Briefly, KYC and AML require financial institutions to conduct due diligence to know the identities of their customers and the nature of their business relationships.[131] Once financial institutions "know their customer," this will help them avoid illicit transactions that may be related to money laundering and terrorism. AML and KYC do not directly impact the vast majority of cryptocurrency investors or users, but both regulations play in important role in the regular use and pubic image of cryptocurrencies.

The introductory chapter covered the brief history of cryptocurrencies, which included Silk Road. The Silk Road online platform allowed for the

> *Peer to Peer /P2P—A connection between two or more computers that allows them to directly share information, files, or other data.*

use of bitcoin for many illegal and immoral activities. That is why the government shut it down. In fact, government regulators learned criminals preferred bitcoin as it was easy to transfer, it could be re-used in peer-to-peer transactions, and perhaps most importantly, it allowed bad actors to circumvent banking institutions to complete current and future transactions. These were the very actions that KYC and AML were designed to prevent. This history also initially placed bitcoin in a bad light as many regulators focused on the small percentage of illegal actions related to bitcoin rather than then vast majority of positive global impact of bitcoin, and later other cryptocurrencies.

If regulators in most major countries—especially G7 countries including: Canada, France, Germany, Italy, Japan, United Kingdom, and the United States—declared bitcoin illegal, the price of bitcoin would have decreased, and worse, would have been cut out of a major global marketplace. Balanced, well thought-out regulations, including KYC and AML have created a balance between allowing cryptocurrencies to create their own future in the marketplace while preventing criminals from using the unique qualities of cryptocurrencies for illegal purposes.

THE DEPARTMENT OF TREASURY OFFICE OF FOREIGN ASSETS CONTROL (OFAC) REGULATION OF CRYPTOCURRENCIES

OFAC's mission is to "administer and enforce economic and trade sanctions based on U.S. foreign policy and national security goals against targeted foreign countries and regimes, terrorists, international narcotics traffickers, those engaged in activities related to the proliferation of weapons of mass destruction and other threats to the national security, foreign policy, or economy of the United States. It regulates and enforces economic sanctions in support of U.S. National Security and Foreign Policy."[132] OFAC treats cryptocurrencies as money for essentially the same reasons as FINCEN. In March 2018, OFAC published *OFAC FAQs: Sanction Compliance,* which includes a section of eight questions and answers (Q&As) called, "Questions on Virtual

Currency." Two of the Q&As updated on March 19, 2019, follow (verbatim).

561. How will OFAC use its existing authorities to sanction those who use digital currencies for illicit purposes?

The United States' whole-of-government strategies to combat global threats such as terrorism, transnational organized crime, malicious cyber activity, narcotics trafficking, weapons of mass destruction (WMD) proliferation, and human rights abuses include targeting an array of activities, including the use of digital currencies or other emerging payment systems to conduct proscribed financial transactions and evade U.S. sanctions. The strategies draw from a broad range of tools and authorities to respond to the growing and evolving threat posed by malicious actors using new payment mechanisms. OFAC will use sanctions in the fight against criminal and other malicious actors abusing digital currencies and emerging payment systems as a complement to existing tools, including diplomatic outreach and law enforcement authorities. To strengthen our efforts to combat the illicit use of digital currency transactions under our existing authorities, OFAC may include as identifiers on the (Specially Designated Nationals) *SDN List* specific digital currency addresses associated with blocked persons.[133]

562. How will OFAC identify digital currency-related information on the SDN List?

OFAC may add digital currency addresses to the *SDN List* to alert the public of specific digital currency identifiers associated with a blocked person. OFAC's digital currency address listings are not likely to be exhaustive. Parties who identify digital currency identifiers or wallets that they believe are owned by, or otherwise associated with, an SDN and hold such property should take the necessary steps to block the relevant digital currency and *file a report with OFAC* that includes information about the wallet's or address's ownership, and any other relevant details.[134]

Violations will result in significant civil and criminal penalties. Similar to FINCEN, OFAC provides regulatory oversight so that cryptocurrencies are not used in an illegal manner, thus stabilizing the use and image of cryptocurrencies as a global currency and investment.

SUMMARY

Regulation of cryptocurrencies and ICOs are critical to prevent unscrupulous businesses from taking advantage of investors. When investing in cryptocurrencies and ICOs, make sure to conduct detailed research, consult with experts, and use the SEC website to see if the cryptocurrency or ICO is mentioned. Even if your cryptocurrency or ICO of interest is not mentioned, it is helpful to know about other enforcement actions. If some of the prior enforcement actions look similar to your cryptocurrency investment of interest, it should be a red flag and you should proceed with caution. Do not let FOMO lead to Loss of Money and Opportunity (LOMO).

SUCCESS STRATEGY ACTION ITEMS

1. Keep updated on CFTC, SEC, and Treasury enforcement actions. As covered in this chapter, some companies have created fraudulent entities to serve one purpose—to defraud investors of their money. It is important to keep track of these companies and their business models so you don't fall into their traps.

2. If you are interested in ICOs, https://topicolist.com/ has a list of ICOs in different stages of development. If you decide to invest in ICOs, be sure to use all of the advice in this book, and especially this chapter, to thoroughly research ICOs of interest.

3. Keep updated on new IRS notices, laws, and regulations as they will play a critical role in your tax liability.

4. Consider looking for a CPA or a tax preparer before you start investing in cryptocurrencies. Make sure the CPA or tax preparer has experience in tax liabilities related to cryptocurrencies. Your tax professional can serve as a resource to answer your cryptocurrency tax related questions.

5. Track the action of the Blockchain Caucus as it will have a positive impact on blockchain technology. In attending numerous Blockchain Caucus events or conferences where U.S. representatives or legislative aides were guest speakers, I have found that members of the Blockchain Caucus understand the potential (and challenges) of the blockchain. The Blockchain Caucus wants to make sure elected officials understand the blockchain before they pass legislation that will prevent the blockchain from reaching its potential.

[5]

BLOCKCHAIN BASICS

> *"I'm reasonably confident...that the blockchain will change a great deal of financial practice and exchange. ...Forty years from now, blockchain and all that followed from it will figure more prominently in that story than will bitcoin."* [135]
>
> —*Larry Summers*
> *Former U.S. Secretary of the Treasury*

THE CREATION OF BITCOIN had a number of powerful consequences, one of which was the invention of the blockchain. Bitcoin, and most other cryptocurrencies, use blockchain technology to verify the cryptocurrency transactions, but what if blockchain technology can be used to verify other business transactions? Blockchain technology could eventually revolutionize business. A good way to predict the use of blockchain technology twenty years from now is to look back on the invention of the internet.

The first version of the internet, Advanced Research Projects Agency Network (ARPANET), invented in 1983, obtained commercial use when it became the World Wide Web in 1990. [136]

Even in the first decade of commercial use, the general public could not predict the importance of the internet in our daily personal and business activities. The same prediction applies to blockchain technology. Many blockchain experts believe the blockchain figures more prominently into business applications than bitcoin. This and following chapters of *Blockchain or Die* cover blockchain basics, the evolution of the blockchain, and blockchain business applications. But first, please allow me to share my blockchain journey.

MY BLOCKCHAIN JOURNEY

My research and investment in cryptocurrencies helped me realize the impressive business applications of blockchain technology. The introduction of cryptocurrencies and blockchain technology, especially Ethereum, represented two "once in a lifetime" opportunities at the same time.

I quickly learned that blockchain technology was more complicated and far reaching than cryptocurrencies. I also learned that blockchain technology was not as far along as bitcoin and other established cryptocurrencies. As covered in the following sections, the vast majority of companies are still researching the specific application of blockchain technology to their companies and industries.

INDUSTRIES USING, OR STARTING TO USE, BLOCKCHAIN TECHNOLOGY

When corporations learned about bitcoin (and more importantly blockchain technology) and they realized blockchain technology had the potential to solve some of their most difficult business challenges, they developed separate blockchains for their own business applications.

Bitcoin and cryptocurrency supporters would argue that once the blockchain is separated from cryptocurrencies, it is no longer a "blockchain." As a result, some corporations refer to blockchain technology as a "distributed ledger technology." For consistency, this book will continue to use the terms "blockchains" or "blockchain technology."

Fortune.com's article, "Blockchain Is Pumping New Life Into Old-School Companies Like IBM and Visa," reported the market for blockchain related products was $242 million in 2016 and would reach $7.7 billion in 2022.[137] Fortune 500 companies like IBM, Visa, Microsoft, and Oracle are heavily investing in blockchain technology and expect major dividends from their investment. In the spirit of the Nakamoto's original concept of a peer-to-peer community, IBM offers a free trial of the blockchain on their cloud, and is a major force behind the Hyperledger Consortium, a non-profit open source project.

Hyperledger.com states "Hyperledger is an open source collaborative effort created to advance cross-industry blockchain technologies. It is a global collaboration, hosted by The Linux Foundation, including leaders in finance, banking, IoT (internet of things), supply chain manufacturing, and technology."[138] Hyperledger is helping companies with their blockchain efforts by providing online access to coding and instructions to build their first network. This chapter will discuss the blockchain starting with the creator of Ethereum: Vitalik Buterin.

THE NINETEEN-YEAR-OLD THAT FOREVER CHANGED BUSINESS

According to BitcoinMagazine.com, Vitalik Buterin, a nineteen-year-old Russian-born programmer from Toronto, Canada, created Ethereum—a platform that allows developers to build blockchain applications.[139] Buterin was inspired by the challenges he faced trying to build applications on the bitcoin blockchain. Buterin believed the potential of blockchain technology was not limited to financial applications and designed a blockchain that could support more common applications.

In 2011, Vitalik Buterin co-founded *Bitcoin Magazine* and wrote many articles explaining his views on the digital currency's future. Unlike Satoshi Nakamoto, Vitalik Buterin has been very involved in the Ethereum and blockchain communities. In addition to founding *Bitcoin Magazine*, Buterin has given many speeches on Ethereum and blockchain technology. When people say they cannot trust or invest in bitcoin because they do not know the identity of Satoshi Nakamoto, that argument does not hold up with Ethereum. I tell them that if the identity of a cryptocurrency creator is that important, consider investing in Ethereum or other cryptocurrencies, especially given that the blockchain is a trust-based system. Similar to the recommendation to read the Bitcoin White Paper, the best way to learn blockchain fundamentals is to read the Ethereum White paper.

THE WHITE PAPER THAT MADE BLOCKCHAINS A BUSINESS MODEL

In 2013, Vitalik Buterin sent out a white paper entitled, *A Next Generation Smart Contract & Decentralized Application Platform*.[140] Buterin's white paper changed business processes and the world of contractual arrangements. Buterin's paper, published five years after Satoshi's white paper, built off of the bitcoin blockchain technology, and arguably had a broader scope and greater impact on businesses, organizations, and governments.

Similar to the analysis of the Bitcoin White Paper abstract in Chapter 1, an analysis of the Ethereum White Paper abstract is helpful to provide an introduction to Ethereum. The bolded terms in the abstract are very important words for truly understanding Ethereum and blockchain technology. The Ethereum White Paper abstract starts by giving credit to bitcoin for creating the first credible decentralized solution to process and confirm digital currency transactions the sets the stage to use the same technology for more than just money. This is where Ethereum comes in. The relevant part of the Ethereum White Paper states:

> "*Commonly cited applications include using on-blockchain digital assets to represent custom currencies and financial instruments ('colored coins'), the ownership of an underlying physical device (**smart property**), non-fungible assets such as domain names ('Namecoin') as well as more advanced applications such as decentralized exchange, financial derivatives, peer-to-peer gambling, and on-blockchain identity and reputation systems. Another important area of inquiry is '**smart contracts**'—systems that automatically move digital assets according to arbitrary pre-specified rules. For example, one might have a treasury contract of the form 'A can withdraw up to X currency units per day, B can withdraw up to Y per day, A and B together can withdraw anything, and A can shut off B's ability to withdraw.' The logical extension of this is **decentralized autonomous organizations** (DAOs)—long-term smart contracts that contain the assets and encode the bylaws of an entire organization.*"[141]

Ethereum evolved the bitcoin blockchain beyond finance to include general technology and business applications by creating a program language to create smart contracts. These smart con-

tracts can be used to create a business and technology structure
for virtually every business scenario. In short, Ethereum created
building blocks to build blockchain applications. Chapter 4 already discussed ICOs, which were designed based on the functionality of the Ethereum blockchain. This chapter will explain
the many applications and business tools including: smart contracts, decentralized autonomous organizations, and decentralized applications.

BLOCKCHAIN DEFINITIONS YOU NEED TO KNOW

Many of the terms in the Bitcoin White Paper also apply to the
Ethereum White Paper and blockchain technology. For example,
the terms "blockchain" and "smart contracts" are in the Bitcoin
White Paper and are reintroduced in the Ethereum White Paper.[142] The Ethereum White Paper introduces new terms as well,
including "smart property" and "decentralized autonomous organizations."[143] As a reminder, I will list the definitions for
"blockchain," "smart contracts," and "Decentralized Autonomous
Organizations" below, along with the new terms from the
Ethereum White Paper.

Blockchain—Blockchain is a public database of all bitcoin
transactions that has ever occurred in bitcoin network. By using
this database, every user is able to find out what amount of
bitcoin has ever belonged to some particular address at a certain
time period.

Decentralized Autonomous Organization (DAO)—A
leaderless organization supported by a network of computers. To
be decentralized, it must have no central location because it is
running on a network of computers. And because there is no single leader and it has its own rules to follow, it is autonomous, or
self-governing.

Smart Contract (also self-executing contract, block-chain contract, or digital contract)—An electronic algorithm that automates the contract execution process in the blockchain. Smart contracts standardize, automate, and exclude divergences in the treatment of the agreement terms by the entered parties.

Smart Property—Property whose ownership is controlled via the bitcoin blockchain (using contracts).[144]

All of these terms will be explained in detail in this chapter.

While the additional terms below are not in the Ethereum White Paper, they are very important to understanding block-chain technology.

Block—A single digital record created within a blockchain. Each block contains a record of the previous block, and when linked together these become the "chain."

Decentralized Application (DApp)—A software application that has its technology running publicly on a network of computers that gives the technology security. DApps are maintained by many individuals instead of by one organization. A hacker cannot alter the application's data unless they are able to get access into nearly all of the network's computers and adjust it there.

Gas—A small amount of Ethereum paid to people who use their computers to record transactions and do other software actions.

Similar to the cryptocurrency terminology in Chapter 1, it is important to understand blockchain terminology. It may be more important given blockchains have the potential to impact every industry. The next section shows blockchain basic properties.

BLOCKCHAIN BASIC PROPERTIES

The above definitions describe individual components of the blockchain, but a condensed description of the basic properties of the blockchain shows the blockchain's impressive potential.

1. Decentralized Systems—A system where elements are spread out with decisions made from many points, and independence is preserved across the network. The diagram in the "Different Types of Blockchains" section below illustrates decentralized systems.

2. Distributed Ledger—A system of independent computers/nodes simultaneously recording data while identical copies of the recording are kept by each computer/node.

3. Cryptographically Secured—The process of making ordinary information unreadable.

4. Mining Validates Transactions—The process of using computer power to solve a complex math problem presented by the crypto system, reviewing and verifying information, and creating a new recording to be added to the blockchain. The first miner to solve the problem creates a new block and receives compensation.

5. Transactions are Immutable—The "immutable property of the blockchain" refers to the Secure Hash Algorithm 256 or SHA 256 as defined in the bitcoin terms in Chapter 1. This algorithm makes guessing the data hidden within the hash virtually impossible and immutable.[145]

It is important to note some of these basic blockchain features, specifically mining and being decentralized, may not be features in more advanced or specific blockchains. As covered later in the chapter, there are other blockchains, hybrid and centralized, that are not decentralized and do not use miners for blockchain operations.

BLOCKCHAIN SCENARIO: TRANSFER OF REAL ESTATE TITLE

Transfer of title in a real estate purchase is an excellent way to illustrate how the blockchain actually works versus our known/common process. The following is a step-by-step comparison of the real estate process and the blockchain process.

PART 1. THE START OF THE PROCESS—REQUEST TITLE INSURANCE

TITLE PROCESS	BLOCKCHAIN PROCESS
The buyer makes the title insurance request to ensure they are receiving clear title.	The request is put into the blockchain to start the process to ensure the transaction is valid.

PART 2. THE BLOCKCHAIN STARTS WORKING—SEARCH THE RECORDS

TITLE PROCESS	BLOCKCHAIN PROCESS
The title company researches the name, tax, and property information and examines it to determine if there are any defects in title.	The miners start to hash the data to determine if all of the transactions since the first transaction are valid.
Exposing hidden defects (unpaid taxes, encroachments, incorrect surveys, etc.). If the title is clear of defects, also known as encumbrances, the title insurance company will issue an insurance policy against the title.	If all of the transactions are valid, including the current transaction, the transactions are cleared for approval.

PART 3. THE FINAL STAGE—PREMIUM PAYMENT AT CLOSING

TITLE PROCESS	BLOCKCHAIN PROCESS
The property buyer pays the insurance premium.	The miners used the blockchain to complete all of the components in Part 2. The first miner that completes and verifies the transaction is paid.

By comparison, the steps in title insurance process covers the basic steps of the blockchain process and similar to other industries, the title process is making its way on the blockchain. In a Globestreet.com article, the blockchain company Propy is described as "a global real estate market place with decentralized title registry," that became the first company to record a blockchain address on a deed. This transaction took place under a pilot program between Propy and a county in Vermont. This is only the start of the introduction to the blockchain, a basic understanding of blockchain protocols is important to understand how Propy, and other organizations, created their blockchain solution.

BLOCKCHAIN CONSENSUS PROTOCOLS

An interesting way to discuss the logic and intent behind consensus protocols is to discuss the Byzantine Fault Tolerance. The Byzantine Fault Tolerance is based on the article, "The Byzantine Generals Problem."[146] Briefly, the Byzantine Generals problem is when generals in the same army use messengers to deliver orders to other generals for a coordinated attack on an enemy location. The Byzantine Generals must create a system to reach consensus before the attack to ensure victory. Compromising any of the orders could comprise the place to attack, time to retreat, or any other military course of action.[147] How does this military problem relate to a consensus protocol?

Consensus protocols, based on the concept of consensus decision-making, are the engine of blockchain. Consensus protocols are the algorithms that carry out the consensus decision-making process. Without the consensus protocols, the blockchain is basically a database that stores information. Wikipedia defines consensus-decision-making as: "a process in which group members develop, and agree to support a decision in the best interest of the whole." Wikipedia further explains the components of the consensus decision-making process below. These all may sound similar, or even identical, but each one of these components plays an important role in the consensus decision process.

- Agreement Seeking: A consensus decision-making process attempts to generate as much agreement as possible.

- Collaborative: Participants contribute to a shared proposal and shape it into a decision that meets the needs of all of its members, rather than competing for personal preferences.

- Cooperative: Participants in an effective consensus should strive to reach the best possible decision for the group and all of its members, rather than competing for personal preferences.

- Egalitarian: All members of a consensus decision-making body should be afforded, as much as possible, equal input into the process. All members have the opportunity to present, and amend proposals.

- Inclusive: As many stakeholders as possible should be involved in the consensus decision-making process.

- Participatory: The consensus process should actively solicit the input and participation of all decision makers.[148]

There are many different consensus protocols that do the work of the blockchain in different ways. The *Hackermoon* article, "ConsensusPedia: An Encyclopedia of 30 Consensus Algorithms," as well as the white papers from the individual cryptocurrencies, provides a comprehensive comparison of the current protocols.[149] As of the time of publication, the following four protocols are the most common.

PROOF OF WORK

Proof of work is the first and most common consensus protocol. In proof of work, miners compete against each other to complete a transaction on the blockchain and get rewarded. Satoshi started bitcoin with the proof of work consensus and Vitalik used the same protocol for Ethereum. With many miners using CPUs or ASIC (Application-Specific Integrated Circuit) computers running the hash algorithms, proof of work is expensive and uses a lot of energy.

PROOF OF STAKE

Proof of Stake (PoS) was created to address the shortcomings of the proof of work consensus protocol. In proof of stake, validators invest in coins in the system and the percentage or mining is directly related to the amount of coins invested in the system. If validators have 10% in the proof of stake system, they have a 10% chance of mining the next block. Proof of stake uses less energy and it is more expensive to attack for hackers.

DELEGATED PROOF OF STAKE

Delegated proof of stake (DPoS) is similar to PoS in that validators still invest coins, but in DPoS, the validators can elect lead validators who will vote on the transaction on their behalf. Since there are less validators voting, the transaction speed is much faster.

DIFFERENT TYPES OF BLOCKCHAINS

Essentially, there are three types of blockchains: open/permissionless/public, closed/permissioned/private, and hybrid/consortium. It is important to know the different types of blockchains as there are cryptocurrencies based on different types of blockchains. Ethereum.org's definition of the three different types of blockchains follows:

> "Open/Public Blockchain—A blockchain that anyone in the world can read, anyone in the world can send transactions to and expect to see them included if they are valid, and anyone in the world can participate in the **consensus process**—the process for determining what blocks get added to the chain and what the current state is. As a substitute for centralized or quasi-centralized trust, public blockchains are secured by crypto-economics—the combination of economic incentives and cryptographic verification using mechanisms such as proof of work or proof of stake, following a general principle that the degree to which someone can have an influence in the consensus process is proportional to the quantity of economic resources that they can bring to bear. These blockchains are generally considered to be "fully decentralized."

Closed/Permissioned/Private Blockchain—A blockchain where write permissions are kept centralized to one organization. Read permissions may be public or restricted to an arbitrary extent. Likely applications include database management, auditing, and highly sensitive data.

Distributed/Hybrid/Consortium Blockchain—A blockchain where the consensus process is controlled by a pre-selected set of nodes. For example, one might imagine a consortium of fifteen financial institutions, each of which operates a node and of which ten must sign every block in order for the block to be valid. The right to read the blockchain may be public, or restricted to the participants, and there are also hybrid routes such as the root hashes of the blocks being public together with an API that allows members of the public to make a limited number of queries and get back cryptographic proofs of some parts of the blockchain state.[150] These blockchains may be considered "partially decentralized."

An illustration of the three types of blockchains follows.

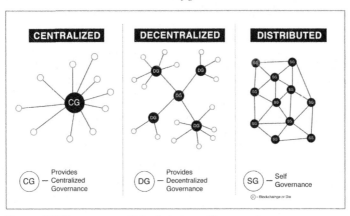

Figure 13 Three types of blockchains—illustrated

In addition to the illustration, the following table in figure 14 provides an introductory side-by-side comparison of all three types of blockchains.

Open/ Permissionless/ Public	Closed/ Permissioned	Hybrid/Consortium
Open read/write	Permissioned read and/or write	Permissioned read and/or write
Slower	Faster	Faster
Anonymous Pseudonymous	Known identities	Known identities
Longer transaction times	Shorter transaction times	Shorter transaction times
Examples	**Examples**	**Examples**
Bitcoin, Ethereum, Dash, Litecoin, Dodgecoin, etc.	Ripple, Hyperledger, Quorum, etc.	XDC, EWF (Energy), B3i (Insurance), Corda, etc.

Figure 14 Comparison of blockchain types

Once a specific kind of blockchain is created, it can be used to resolve many business challenges. Smart contracts are one of the tools created from the blockchain to streamline business practices and solve business challenges.

What are Smart Contracts?

The term "smart contract" was defined earlier in the chapter, but what does smart contract really mean?[151] Traditionally, contracts are legal agreements, but smart contracts are not necessarily legal agreements. Smart contracts are automated, self-verifying, and secure agreements that set the rules and terms for the parties to perform their duties to complete the agreement.[152]

Many articles give examples of smart contract transactions including the common transaction of buying and selling a vehicle. When a seller advertises a car on the blockchain for a certain price and a buyer likes the car and the price, the buyer can use his/her public and private keys to transfer cryptocurrency to the seller to purchase the vehicle. Generally, this is a good example as most people have purchased or sold a vehicle. But there are other considerations when buying and selling a car the blockchain cannot verify or record at this time, including title, insurance, license plates, inspection, etc. All of these are part of the transaction process when buying and selling a car.

Showing how Sia, a blockchain cloud storage company, uses smart contracts may help illustrate how they work. Sia has a similar business model to Dropbox or Google Drive, but Sia does not require physical storage locations because all of the Sia's client data is decentrally stored on the blockchain.[153] Computer owners can "rent out" extra hard drive space on their computer and receive payment in Siacoin. The transaction takes place by smart contract on the blockchain. Siacoin is an example of a smart contract that is currently available to computer owners who can use their spare hard drive space to make a profit. Computers owners do not need to know how to program or code to profit from the blockchain.

Sia is an example of a company using smart contracts to connect computer owners to the blockchain. There are companies solely made up of smart contracts where the smart contract is the primary product and administrative resource for the organization. These organizations are called Decentralized Autonomous Organizations.[154]

DECENTRALIZED AUTONOMOUS ORGANIZATIONS EXPLAINED

Historically, centralized organizations operate in a linear fashion. All of the decisions are made at or near the CEO or executive level and filter down to the employees. The following diagram describes the typical centralized organization.

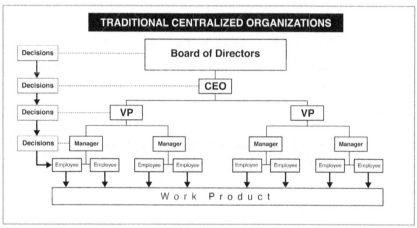

Figure 15 Example of a traditional centralized organization

Decentralized Autonomous Organization (DAO) is another use of blockchain technology that is growing exponentially. But what is a DAO?

DAOs, prominently mentioned in the Ethereum White Paper, are organizations created in the blockchain that use smart contracts to carry out the business of the DAO.[155] As DAOs are new and are still evolving, there are many different definitions for the DAO, but every credible version of the DAO definition has three core components: programmed set of rules, operate autonomously, and use and distribute consensus protocol. InvestinBlockchain.com is one resource that provides the three components:[156]

1. The DAO is a programmed set of rules. This embedded programming creates the smart contract that is the core of the DAO.

2. Once the programming is in place, it is designed to operate autonomously. This means people are not responsible for the day-to-day activities of the DAO, the programming runs all of the activities.

3. The program is run through a distributed consensus protocol. Similar to the solution in the Byzantine Generals problem, decisions regarding the future of the platform are taken by the community of users according to the agreed initial plan.

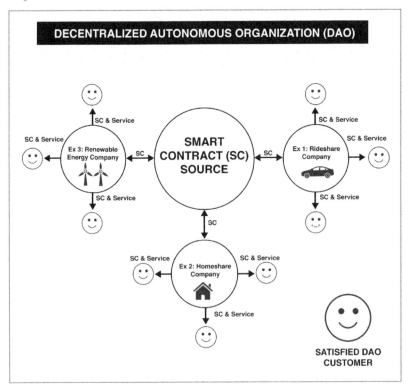

Figure 16 Three DAO examples: ride share, home share, and blockchain microgrid

An Intuitive Approach to DAO

Another way to describe the term Decentralized Autonomous Organization is to break the term into individual words: decentralized, autonomous, and organization. From this perspective, the term "Decentralized Autonomous Organization" is decipherable. Similar to bitcoin and other digital currencies, DAOs are decentralized. As previously discussed, "decentralized" for a currency means the government does not control the currency. What does "decentralized" mean for a corporation? In theory, it means the central authorities (chief executive officer, other senior executives, and a board of directors) are not responsible for the day-to-day operations of the company. That brings us to our second word in DAO: "autonomous."

Autonomous means all of the work of the corporation will be completed by the blockchain. DAOs are operated by a series of smart contracts, not by employees. In theory, once the smart contracts are created, the organization can run on its own on the blockchain. That also means, once the contract is created, the contract cannot be changed and must be completed based on the original specifications. I am of the opinion that an organization cannot be "fully" autonomous. An organization, especially one based on software, needs constant updates and maintenance to remain relevant, but it appears that the day-to-day smart contract activities: operations, procurement, awarding, completion, and payment can be completed in an autonomous manner.

"Organization" is self-explanatory. Wikipedia defines organization as: "an entity comprising multiple people, such as an institution or an association, that has a particular purpose."[157] Since the term is "organization," not "business" or "corporation," DAO can be used by anyone for any purpose. For example, volunteer organizations, churches, and alumni associations can create

DAOs to serve their purpose. The main commonality for all DAOs is the use of smart contracts.

Blockspur.com—A Resource for Ranking Smart Contracts

Similar to CoinMarketCap.com serving as a cryptocurrency resource, Blockspur.com serves as a resource for decentralized apps, smart contracts, and digital goods. Blockspur.com differs from CoinMarketCap.com in that it ranks smart contracts by usage (transactions, users, and value) while CoinMarketCap.com primarily uses market capitalization and market value.

Blockspur displays the metrics associated with each smart contract by number of transactions.

Contracts By Number of Transactions

(For July 2019)

	Change		Address	Name	Symbol	ERC20	Transactions	Uniques	"Revenues" in Ether	"Revenues" in USD
1	N/A		0x4f01a01...			No	301,256	1	0.00 ETH	0.00
2	-1		0x2e0c0db.	Exchange		No	290,452	17,689	67,230.13 ETH	$5,781,791
3	-1		0xf1ceeee...			No	250,639	820	34,972.39 ETH	$3,007,625
4	~4		0x8fdcc30...	CpcToken	CPCT	Yes	147,724	10,396	0.00 ETH	0.00
5	-2		0x06012c8...	CryptoKitties	CK	Yes	139,776	2,418	386.17 ETH	$33,211
6	~926		0x93x6dfE.			No	100,477	3,631	0.01 ETH	$1
7	0		0x8d12a19...	EtherDelta		No	91,653	10,593	26,086.14 ETH	$2,243,408
8	~4		0x8dahd3r...	Tuzy Coin	TUC	Yes	90,009	1,788	0.00 ETH	0.00
9	↓4/		0x14frva9...	MatchingMarket		No	77,747	859	0.00 ETH	0.00
10	~196		0x600a0e...	Registrar		No	72,022	735	83.89 ETH	$7,21b
11	~1328		0xDba3e6e...			No	70,300	5,970	2,562.10 ETH	$220,341

*Assts that are blank (specifically name and symbol) did not have any information listed on blockspur.com.

Figure 17 Smart contract rankings by number of transactions

Smart contracts are the foundation for DApps and DApps facilitate the use of smart contracts. Working together, smart contracts and DApps make the blockchain accessible.

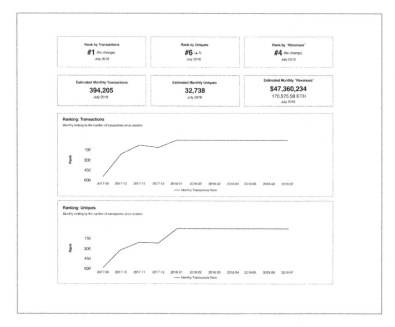

Figure 18 Blockspur number one contract detail

INTRODUCING A NEW ERA IN APPS

As defined earlier in the chapter, DApps are software programs, similar to existing iPhone or Android applications, that provide a user-friendly "technology storefront" for users to connect to businesses. DApps differ from current applications, as they are peer-to-peer networks that directly connect buyers and sellers rather than providing access to a centralized company.

The Ethereum White Paper defines three types of applications that can exist on top of the Ethereum platform.

1. Financial applications that enable users to enter into financial contracts.

2. Semi-financial applications where a financial component is involved but not the sole purpose of the contract.

3. Non-financial applications such as online voting and other uses.[158]

A DApp can be based on any one of these three application categories.

Similar to CoinMarketCap and Blockspur, State of the DApps is a resource for DApp information. State of the DApps is a directory of decentralized applications on the Ethereum blockchain. State of the DApps categorizes and showcase projects built on the Ethereum blockchain. The DApps directory covers different fields such as: health, games, virtual reality, artificial intelligence, education, registries, job markets, tinder for horses (yes an app for horses), and many more.

In April of 2015, State of the DApps reported twenty-five decentralized applications. In July 2019, State of the DApps reported 2,667 decentralized applications, with 30.59k daily users.[159] State of the DApps is a good example of the explosive growth of decentralized applications (see figure 19).

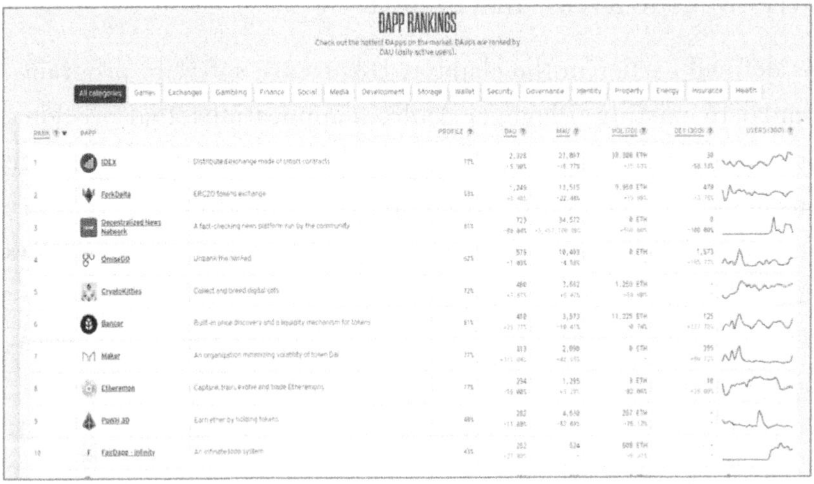

Figure 19 State of the DApps rankings

DAOs in the Age of Uber

Everyone with a smart phone uses apps. Touch the app once and you are online booking flights, buying clothes, investing in stocks, and even swiping right to find a date. Apps even help us travel and find places to stay. Of course, this is a reference to Uber and Airbnb. Most urban and suburban dwellers know about Uber. Uber is clearly a global company with a global brand. How is the Uber comparison relevant to DAOs? The Uber user's request for a ride is submitted through the centralized Uber app, which has user credit card information and personal information. The Uber driver picks up and drops off the rider, the rider's credit card is automatically charged, the rider pays Uber, and Uber pays the driver, and the transaction is complete.

Current rideshare services, such as Uber and Lyft, require all drivers and riders to transact through the corporation. This middleman sets ride prices, collects a percentage of the fees paid to the driver, and owns all data for every rider. When the middleman gets hacked, everyone's data is stolen. Meanwhile, drivers are compensated as independent contractors, but they are treated like employees with no autonomy. It's time for something better. Enter decentralized ride sharing.

In decentralized ride sharing, rides are requested through a decentralized peer-to-peer application. As the transaction is decentralized the rider's credit card and personal information is not in a centralized location. Once again, centralized storage of rider information is a target for hackers. *CBS* reports a hack at Uber resulted in the loss of names, addresses, mobile numbers, and emails of fifty million Uber customers as well as seven million drivers.[160]

Riders pay decentralized ride share in cryptocurrencies, which protects their personal and credit card information. A decentralized ridesharing company will not employ hundreds of employees to support the company, and when cars become fully automated, they will not need to use drivers. Does this sound far-fetched? Ridecoin and La`Zooz are two ride sharing companies that will use the blockchain.

Ridecoin evolves ridesharing by replacing the middleman with the blockchain. Ridecoin's peer-to-peer network will allow drivers and passengers to control the ride sharing transactions and the data. Ridecoin passengers pay less while having more control and the money and data are not in the hands of a centralized company.[161] Finally, drivers are their own bosses and don't have to follow company policies or rely on incentives that do not solely focus on rewarding the driver.

La`Zooz, another DAO ride sharing company, is a decentralized transportation platform that utilizes empty vehicle seats to create smart transportation solutions. According to La`Zooz.org, La`Zooz will "synchronize empty seats with transportation needs in real-time, matching like-minded people to create a great ride-sharing experience for a 'Fair-Fair.'" Uber created the ridesharing experience, and Ridecoin and La`Zooz are using the blockchain to evolve the ride sharing experience.[162]

Airbnb is another global company with a global brand that disrupted the travel industry. Travelers went from staying in hotels to staying in people's homes, and in 2018, *Fortune* reported Airbnb was worth $38 billion. But there are positive blockchain disruptions on the horizon, enter Bee Global and CryptoBnB.

Similar to Ridecoin and La`Zooz evolving the ridesharing business, Bee Global and CryptoBnB are evolving the homesharing business. Both companies use blockchain technology to match homeowners with temporary tenants, removing the centralized intermediary and reducing fees.[163 164] Under the DAO model,

BeeGlobal and CryptoBnB will operate with minimal staff and overhead passing the savings to the homeowner and temporary tenant. These examples show how the blockchain will impact everyday life.

"HOW DO I GET ON THE BLOCKCHAIN?"

During blockchain speeches and trainings, I am frequently asked: "Eric, this is great information and I am interested in the blockchain, so how do I get on the blockchain?"

Most people learning about blockchain technology for the first time are under the impression they need programming experience to access a blockchain. The following questions help put the use of blockchains in perspective.

My first question to them is if they have a personal or professional website or social media page. So far, everyone has confirmed they have a website or social media page.

My second question is if they know HMTL or any other code to create their website or social media pages, and so far no one stated they coded their own website.

Finally, I ask how they created their website or social media page, and most provide one of the three following responses:

1. They hired a personal webpage designer.
2. They used an online webpage or social media platform.
3. They used a social media page rather than a website.

Blockchains work the same way. Initially use of blockchain technology required programming experience, but years later many companies created products and services to provide easy access to blockchains that are readily available to the average user. These products and services will be discussed in detail in the next chapter.

SUMMARY

This chapter provided the basics for blockchains including: definitions, protocols, types of blockchains, smart contracts, and
DAOs. For most readers, this is the first exposure to this information, but the term "blockchain technology" is starting to make
its way into everyday conversations.

During professional social events, when I introduce myself as
the author of *Blockchain or Die*, I am usually asked, "What is the
blockchain"? At that point all heads turn to me and I am on the
spot to impart words of wisdom to explain blockchain technology.
In preparation for the many times I will be asked this question in
the future, I constructed a thirty-second blockchain elevator
speech to explain the fundamental use of blockchains. The best
part is that all of the components in the elevator speech are covered in this chapter.

*If you understand how to purchase a home, spend a dollar, and bal-
ance your checkbook, you basically know how blockchain technology works. Let's start with home ownership. When the buyer
purchases a home from the seller, the title company researches the
property records looking at every transaction leading up to the
current purchase for the property to verify there are no problems
and that the seller can pass clear title to the purchaser.*

*Now, imagine that property is this one-dollar bill. A blockchain
will search and verify every time this one-dollar bill was used to
buy something, up to and until I buy something off of you.*

*The current centralized financial system is primarily focused on
the latest transaction because they store all of their records on
centralized databases that no one else can search or verify. Now,
imagine you can see every time everyone uses this one-dollar bill
and every use is recorded on a ledger similar to your checkbook*

ledger. The checkbook ledger is a record of all of the transactions associated with your account; the only difference is this is a decentralized checkbook ledger for a global cryptocurrency that everyone can see and verify. Combined, these three transactions describe the basics of blockchain technology.

This chapter contained a lot of information on blockchain technology. The basic concept of blockchain technology is not complicated, but the use of public blockchains is a dramatic change in our perception of money and technology. In addition, blockchain technology opens the doors to so many possibilities that it is difficult to explore them all in one chapter.

As advised in previous chapters, the best approach to better understanding blockchain technology is to use the steps in the success action items to actually use blockchains. If you need help understanding any of the blockchains you are using, it is helpful to turn to blogs or join a blockchain organization. It has been my experience that both entities are very welcoming and helpful, especially to newcomers—after all, bitcoin started as a grassroots movement and volunteers helped get it up and running.

SUCCESS STRATEGY ACTION ITEMS

1. Read the Ethereum White Paper—*A Next Generation Smart Contract & Decentralized Application Platform*—at least three times. The definitions in this chapter should help explain some of the key concepts in the white paper.
2. Read "The Byzantine Generals Problem" by Leslie Lamport, Robert Shostak, and Marshall Pease. This article creates a parallel between military and computer system consensus problems and presents several solutions.

3. Create an Ethereum wallet. You can visit www.Ethereum .org and create a free Ethereum wallet. Similar to creating a bitcoin wallet, discussed in Chapter 1, the exercise to create an Ethereum wallet is another way to gain exposure to blockchain resources.

4. Create a smart contract. The best way to fully understand the smart contracting process is to create a smart contract. You can create a smart contract on a number of different platforms, including: Ethereum, Ethereum Classic, Iota, NEM, EOS, and Cardano. Be sure you have enough memory on your hard drive to download a number of programs including a wallet to create a smart contract.

5. Research DAOs in your industry. Although DAOs are in the early phases of use, many DAOs are developing business solutions.

6. Visit State of DApps to see if there are any decentralized applications in your industry or related to a blockchain business idea you are currently considering.

7. Visit Blockspur.com to understand the progress of smart contracts. Blockchains will continue to grow as the number of smart contracts increases in use and dollar amount.

8. Research the potential of using blockchain cloud storage companies as a source of blockchain revenue. This chapter referenced two blockchain storage companies, but there are other blockchain storage companies that may have different blockchain offerings.

[6]

BLOCKCHAIN BUSINESS AND GOVERNMENT APPLICATIONS

"If we remember, fifteen years ago if you were doing anything on the internet you were going to make millions. I think it could be the same with bitcoin."[165]

—*Jared Kenna, Co-Founder, Tradehill*

ONE OF THE REASONS the title of this book is *Blockchain or Die* is because businesses can still get a head start on their competitors and make, or save, millions with blockchain technology. Now that blockchain fundamentals have been established, this chapter focuses on the business applications of blockchain technology.

BLOCKCHAIN WINNERS AND LOSERS

As with any new technology, especially one as groundbreaking as blockchain technology, winners and losers are inevitable. For example, the internet and social media profoundly impacted many industries, including publishers (newspapers, magazines, and books), the music industry, television, movies, advertising, public relations, hospitality, and personal transportation. Similarly, blockchain technology will impact virtually every industry, as such, the first set of winners will be the businesses that effectively incorporate and monetize blockchain technology into their business model.

The blockchain's verifiable and transparent nature threatens the status quo. The decentralized nature of blockchain technology provides greater access to information previously transacted, stored, and transferred through intermediaries. Some businesses, especially businesses that act as intermediaries, will be required to dramatically change their business model to stay in business. These are the businesses with the greatest risk of becoming a "blockchain loser."

INDUSTRIES USING BLOCKCHAIN TECHNOLOGY

Blockchain technology has been a positive and powerful disruptor to many industries. It is insightful to show the industries impacted by blockchain technology, including in alphabetical order:

- Agriculture
- Artificial Intelligence/Internet of Things
- Cloud Storage
- Energy Management
- Financial Services
- Food Industry
- Gambling Industry
- Government and Public Records
- Health Care
- Higher Education
- Human Resources
- Insurance Industry
- Intellectual Property

- Legal
- Media and Entertainment
- Real Estate
- Retail Management
- Publishing
- Supply Chain Management
- Transportation

This list is just the start of the impacted industries. For example, Techcrunch.com reports Wal-Mart is using the IBM Blockchain Solution to create an IBM Food Trust Solution that will digitize the food supply chain process. Using the blockchain to track the source of food will dramatically reduce the tracking time frame from seven days to an estimated 2.2 seconds.[166] This Wal-Mart example incorporates the following industries: agriculture, food industry, supply chain, and transportation. As blockchain technology matures and becomes more useful and affordable, many more industries will discover creative uses for blockchain technology.

This chapter will cover the impact of blockchain technology in healthcare, cloud storage, and human resources industries because the average person has constant interaction with these industries. For example, most people:

- Have health insurance and one or more doctors,
- Own a computer or a smartphone or use cloud storage, or
- Have a job with a personnel folder that contains their personal information or are looking for a job.

The blockchain analysis will follow in this format: first, my personal challenges in healthcare, cloud storage, and human resources; second, the industry challenges and their solutions; and third, blockchain legislation, including current use cases. In addition, this chapter will discuss the impact of blockchain technology on employment verification and diversity. Finally, this chapter will show how U.S. state legislatures are addressing blockchain technology.

THE HEALTHCARE INDUSTRY—CHALLENGES & SOLUTIONS

PERSONAL HEALTHCARE CHALLENGES

During a routine doctor's appointment, the office staff asked me to fill out a new medical information form. As with all medical forms, I was required to include my social security number. When I asked the office staff why they were requiring me to fill out forms with my social security number again, I was advised some of the doctors left the practice and started their own practice. As a result, my doctor was requiring all of her patients to complete new medical information forms. *Great,* I thought, *my social security number is going to be in two different systems with double vulnerability to hackers, not to mention in two hard copies in two places as well.* I was not happy, but I had to fill out the medical form to continue with my appointment. The answer to this problem is in the blockchain healthcare solutions below; but first let's look at another medical challenge I encountered.

While on travel to Colombo, Sri Lanka, one of my medical conditions flared up. I was miserable and it was affecting my work, so I decided to seek local medical attention. The medical doctor was forced to rely on my summary of my medical condition for his diagnosis. Fortunately, my condition is not serious and together, the doctor and I were able to create a medical solution that solved the problem until I returned to the U.S. But what if I had a very serious medical condition? What if the doctor needed detailed medical records, including doctor notes, prescriptions, and information to address my medical issue? What if my medical issue required emergency surgery? How could the doctor obtain my medical information on short notice? How can the blockchain solve this challenge?

My healthcare challenges are not unique; in fact they are rather common. Every medical form requests your social security number, which means your identity is exposed and at the mercy of the doctor's office. Currently, your medical records are either kept on your doctor's computer or a closed health network that is controlled by a combination of administrators or doctors. Simply put, you are not in direct control of your medical records; your doctor or health network controls your medical record. If someone asks for your medical record, you probably have to submit a form to your doctor to release your record; you may even have to pay a fee. Medical record requests submitted during business hours may take a day or more to complete and requests submitted after office hours may take longer. Remember my medical flare up in Sri Lanka? What if it was a medical emergency and I had an urgent need for my medical record? The request for my medical records would have probably delayed my treatment.

HEALTHCARE INDUSTRY CHALLENGES

The health care industry is facing numerous challenges, including ineffective electronic medical record (EMR) systems, accessibility, security, and privacy issues.

EMR specific challenges include:

- Access to medical records is an issue as there are many different EMR systems that operate off of proprietary platforms and the vast majority of EMR systems do not talk to each other.
- EMR systems are not readily accessible to the patient or other doctors, even in the case of an emergency. Patient

portals have increased patient access to medical records, but when patients have numerous patient portals, they also have to access numerous portals to obtain their medical records.

- EMR systems do not directly benefit the patient; they administratively benefit the doctor's office by creating administrative convenience and the EMR company by creating profits from EMR sales and updates.

- EMR systems contain patient's private information (name, social security number, date of birth, and address) making them an attractive target for hackers.[167]

General health care challenges include:

- Lengthy administrative responsibilities.
- Sharing of information between doctor and patient.
- Massive amounts of data related to research, clinical trials, and billing.
- Improper payments. For example, in 2016, Federal health officials made more than $16 billion in improper payments to private Medicare Advantage health plans and when the payment analysis includes overpayments to standard Medicare programs, the number explodes to nearly $60 billion.[168]
- Medicare fraud. Medicare scammers steal 60 billion dollars a year. [169]
- Hacking health care systems. According to www.data breachtoday.com, Anthem Inc., the U.S.'s largest health care system, was hacked and compromised the personal information of 79 million people. Anthem Inc. settled the hacking related lawsuits for $115 million, the largest settlement for a data breach.[170]

- Breach in patient records. *Protenus Breach Barometer Report* shows that in 2017, 56 million patient records were hacked.[171]

Independently, each one of these challenges is a serious problem for the healthcare industry. Together they represent an industry in great need of reform to provide adequate medical services. Enter blockchain technology.

HEALTHCARE INDUSTRY SOLUTIONS

Blockchain technology will revolutionize medical records. Medical records on the blockchain are immediately accessible. You can use a device, an internet connection, and your public and private key to access your records anytime from anywhere in the world. The impact of the blockchain goes beyond medical administration; it will change the practice of medicine.

If a patient is dealing with a chronic or serious health condition and has multiple doctors, all doctors with the public key will have access to the most recent medical reports. Multiple doctors can even provide feedback, comments, or recommendations that may be critical to the big picture for the patient's medical care. Medical referrals can be instantaneous. Prescriptions can be ordered on the blockchain; yep, no more prescription pads.

As previously discussed, the blockchain does not contain any personally identifiable information. No names, addresses, dates of birth, email addresses, phone numbers, and most importantly, no social security numbers! That's right, the days of social security numbers on medical documentation will hopefully be a relic of the analog age. Patients will use their public key and private key to provide personal and medical information.

Blockchain healthcare solutions are already in progress. According to the *IBM Institute for Business Value* abstract, "Healthcare Rallies for Blockchain":

> "The latest IBM Institute for Business Value blockchain study surveyed 200 healthcare executives—both payers and providers—in sixteen countries. We found that sixteen percent aren't just experimenting; they expect to have a commercial blockchain solution at scale in 2017. These Trailblazers are leading the charge with real-world blockchain applications that they expect to take down the frictions that hold them back. They're keenly focused on accessing new and trusted information, which they can keep secure, as well as entering new markets."

Blockchain health care solutions will not work unless they are usable by the average user. DApps displays some of the blockchain solutions that will provide access to the average user.

HEALTHCARE DAPPS

DApps were introduced and explained in Chapter 4. As of July 23, 2019, here are the State of the DApps top ten health care DApps with a brief description.

1. Actifit—Rewarding your everyday activity
2. PUML—Fitness for data, data for rewards
3. VChain—Anti fake by blockchain technology
4. Elysium—Securely store and share patient health records
5. Iku—Biotech redefined
6. Healthereum—Loyalty rewards platform for patients
7. Wowmart—Motivation, promise, reward
8. Molecule—The next evolution in drug development
9. Iryo—Redefining Global Healthcare
10. Enda—Enjoy your DNA[172]

The presence of healthcare DApps shows health care DApps are positioned to grow and add value to the health care industry.

HEALTHCARE INDUSTRY SUMMARY

Health care may be the most important industry to evolve the blockchain as lives are literally at stake and virtually every industry serves as a supporting industry to health care. For example, every industry listed earlier in the chapter has a potential impact on health care, even gambling. How does gambling factor into healthcare? Excessive or abusive gambling is a well-known healthcare issue. If a blockchain business can design and monetize a blockchain product to address excessive or abusive gambling, they create a solution to manage this health care problem.

CLOUD STORAGE INDUSTRY CHALLENGES & SOLUTIONS

MY CLOUD STORAGE CHALLENGE

When I started my company, Better ME Better WE, I created and collected a lot of research, pictures, and videos. My hard drive quickly ran out of memory that slowed my computer down, so I started to use an external hard drive. Although my fellow entrepreneurs recommended using cloud storage, I never felt my company information was safe on "the cloud." When a number of large cloud storage companies were hacked, my concerns were confirmed and I never used cloud storage. Enter blockchain technology.

CLOUD STORAGE INDUSTRY CHALLENGES

Cloud storage companies provide storage on centralized hard drives where the information can be accessed anytime from anywhere. Cloud storage requires a physical cloud storage facility and a large investment in IT security as the data on the cloud is an attractive target to hackers. Cloud storage companies pass both of these large expenses to their customers. Information centralization also means the information is vulnerable to hackers. Hackers are attracted to cloud storage companies since all of the information is in one place for hackers to steal, post on the web, or even hold for ransom. Blog.Storagecraft.com lists seven of the largest cloud storage hacks that affected millions of people:

1. Microsoft—In 2010, hackers accessed employee contact info, but Microsoft allegedly responded within two hours of the breach.
2. Dropbox—In 2012, hackers accessed 68 million user accounts including email passwords and passwords that were reportedly sold on the dark web.
3. National Electoral Institute of Mexico—In 2016, the system was breached resulting in illegal access to 93 million voter registration records.
4. LinkedIn—In 2012, hackers stole six million user passwords that were published on a Russian forum.
5. Home Depot—In 2014, hackers exposed Home Depot's point of sale terminals at the self-checkout lines impacting 56 million credit card users.
6. Apple iCloud—In August 2014, Apple's iCloud service was hacked exposing many private photos leaked online.
7. Yahoo—In December 2015, Yahoo confirmed that a 2014 attack compromised more than one billion accounts.[173]

These examples of cloud storage security breaches show the need for decentralized blockchain cloud storage.

Cloud Storage Industry Solutions

Essentially, blockchain cloud storage provides greater data security and privacy for a lower price. In cloud storage, the blockchain cloud storage company pays for spare hard drive space. In the "blockchain cloud," files are sharded, broken apart, and distributed to numerous nodes. In sharding, a large quantity of information is broken into smaller pieces and distributed it to multiple computers. Essentially, this makes the data or computer programming more manageable.[174]

The private key is the critical component of the blockchain cloud storage. Since the files are imprinted with the same private key, the individual node cannot access the information in the file. In addition, since the files have been broken apart, in the unlikely event a hacker breaks the private key and accesses the file, they will receive an incomplete file. The information they receive from the hack will be fairly useless.

> *Private Key—A string of random letters and numbers known only by the owner that allows them to spend their cryptocurrency.*

Storj is one of the largest blockchain cloud storage companies. Storj's White Paper, "Storj: A Decentralized Cloud Storage Network Framework," provides a solution to cloud storage industry problems:

> "A peer-to-peer cloud storage network implementing client-side encryption would allow users to transfer and share data without reliance on a third party storage provider. The removal of central controls would mitigate most traditional data failures and outages, as well as significantly increase security, privacy, and data control. Peer-to-peer networks are generally unfeasible for production storage systems, as data availability is a

function of popularity, rather than utility. We propose a solution in the form of a challenge-response verification system coupled with direct payments. In this way we can periodically check data integrity, and offer rewards to peers maintaining data."[175]

Even though the Storj is in a different industry than bitcoin, the Storj White Paper uses some of the same terms ("peer-to-peer" and "nodes") as the Bitcoin White Paper. Storj is using blockchain technology to disrupt the cloud-sharing business.

Sia is another blockchain cloud storage company with a similar business model to Storj. A simplified example of how the files are sharded and stored on a blockchain follows:

Before the files are sharded:

File Section 1: AAAAAAAA

File Section 2: BBBBBBBB

File Section 3: CCCCCCCC

File Section 4: DDDDDDDD

After the files are sharded they are stored on nodes as fragments:

Node 1: AAAACCCC

Node 2: BBBBCCCC

Node 3: BBBBDDDD

Node 4: AAAADDDD

Node—Any computer, phone, or any other computing device that can receive, transmit, and/or contribute to a blockchain is a node.

Even if a hacker breaks into any of the storage centers, they will not have access to the whole file; only useless fragments of information. Cloud storage innovated data storage, blockchain cloud storage is evolving cloud storage. When the blockchain cloud storage users want to retrieve the data, the files will be pulled from the nodes and reconstituted to the original file format.

CLOUD STORAGE INDUSTRY SUMMARY

Cloud storage has the potential to impact every industry. The creation, transfer, and secure storage of data is a business factor for every business, especially if the business transacts or stores sensitive data. One important difference between blockchain storage companies and blockchain companies in other industries is that multiple blockchain companies are currently providing blockchain cloud storage solutions.

HUMAN RESOURCE INDUSTRY CHALLENGES & SOLUTIONS

MY HUMAN RESOURCES CHALLENGES

In recent years, my personal information was hacked from one of my former employers. That means hackers have my personal information and can use it whenever they want. My former employer notified the hacking victims of the data breach and provided credit-monitoring services at no charge. In this unfortunate situation, my former employer, like many other employers, provided the best possible resources under the circumstances. My personal information would be safe if I was not required to give my employer my personal information to pass background checks and obtain financial and medical benefits.

HUMAN RESOURCES INDUSTRY CHALLENGES

As discussed in my personal challenges, Human Resources departments are responsible for a large amount of financial and personal information of employees and former employees. Protection of this sensitive data is costly and time consuming and

often requires large IT departments. Human Resources departments are excellent candidates for blockchain solutions especially
in information storage, recruiting, pre-employment verification,
and payroll/finance.

Human resources departments store all of the employee's personal information including name, address, social security number, and probably direct deposit information. Similar to the
discussion in cloud storage, sources of centralized information
are attractive targets for hackers. Vulnerability of employee's
information is a major human resources challenge as any hack-
ings personally impact the employee.

When recruiting for candidates, many corporations rely on intermediaries; specifically search firms, to source new employees,
especially in higher-level positions. Many human resources (HR)
departments do not have time or staff to source candidates domestically or internationally. As a result, companies use external
recruiters or search firms to locate and even screen candidates.
The use of recruiting firms as intermediaries is an additional ex-
pense for companies.

Pre-employment verification is another HR challenge, as most
companies require applicants to verify pre-employment and education before the hiring process is finalized. The applicant is required to submit the following for employer verification:
education; prior employment; and criminal, credit, and financial
history. This is a lot of information and the quality and accuracy
of the applicant's background information may determine the
length and complexity of the background check process. There
may be times when an educational institution or a former employer is no longer in business. The lack of availability of these
background sources adds additional challenges to the already difficult pre-employment verification.

Once the successful candidate is chosen, many corporations require the successful applicant to complete a background check, which may include federal and state criminal background checks as well as financial background checks. Background checks can take weeks, months, or even a year if the position is a highly sensitive government position requiring security clearance. Lengthy background checks means the position is filled, but none of the work will be completed by the new hire until they start in the position. This leaves the employer with a vacant position until the background check is complete and the employee starts work.

Payroll is another challenge businesses face every pay period. Since payroll is subject to state and federal laws, companies have to ensure their payroll system is compliant. Payroll systems are centralized and expensive to maintain, as they require adequate security to protect against hackers and unauthorized attempts to access the employee's data. Multi-national corporations have additional payroll challenges as they have to run payroll in different countries requiring compliance with local laws and converting compensation to the local fiat currency.

HUMAN RESOURCES INDUSTRY SOLUTIONS

Blockchain technology can provide efficiencies in employee information storage, pre-employment verification, the recruiting and hiring process, and payroll systems. When an employer is looking for applicants, they can search blockchains for qualified applicants. How is this different from searching résumé databases or websites like LinkedIn? The first part, the search for an applicant, is very similar to the current online search. Employers can enter keywords into job search websites to find qualified candidates. The second part, the verification of the information on the résumé, is where blockchains provide an improvement over the current technology.

Blockchains can bring additional security to sensitive employee data similar to the additional security it brings to cloud storage. Since the employee data will be decentralized on the employer's blockchain; hackers will face greater challenges to access employee data providing greater access for less money. The method of verification will depend on the type of blockchain: open, closed or consortium.

Instant verification of information on blockchains dramatically shortens, and may eventually eliminate, lengthy background check processes. Human resources departments will not have to contact past employers to verify prior employment and work experience. Prospective employees can upload their employment history to their blockchain and give the prospective employer access to their blockchain to instantly verify prior employment history. Background verification will be more effective as more individuals and employers put employment history on blockchains.

I predict the same principles of employment verification will apply in blockchains; specifically, posting the following information on blockchains: name, position, date hired, date of departure, and perhaps salary. Essentially, this is the information employers provide in the current employment verification process.

Employers can hire blockchain companies to incorporate a blockchain into their business model, which will save the employer time, money, and resources.

VERIFICATION OF CANDIDATES EDUCATION

Educational institutions will play an important role in blockchain technology by creating policies and practices to upload diplomas, degrees, or certificates to their blockchains. Once uploaded, candidates can give employers access to their blockchain to instantly verify their educational achievement. Massachusetts Institute of Technology (MIT) is at the forefront of educational verification. According to *MIT News*:

"MIT is one of the first schools to issue "recipient owner virtual credentials [which] enables students to quickly and easily get a verifiable, tamper-proof version of their diploma that they can share with employers, schools, family, and friends. To ensure the security of the diploma, the pilot utilizes the same blockchain technology that powers the digital currency bitcoin."[176]

Validation of education does not only apply to educational institutions; similarly, continuing education programs, certifications, licenses, and other training related programs can upload course completion to their blockchains for instant verification by an employer. Blockcerts is an example of a blockchain company that places records on blockchains. Blockcerts.com provides a summary of their services:

"Blockcerts is an open standard for building apps that issue and verify blockchain-based official records. These may include certificates for civic records, academic credentials, professional licenses, workforce development, and more."[177]

Once blockchain certification companies like Blockcerts become more common and corporations use blockchain certification companies to verify candidates and applicants, blockchains will become commonplace in the HR solution. However, it will probably take years for blockchains to have enough verified employer information from employers to render the information useful. As with all of the other blockchain business applications, the more it is used, the faster it will be useful.

THE BLOCKCHAIN IN THE JOB SEARCH PROCESS—SOLUTIONS

One of the most important tasks for a jobseeker is to attract potential employers by posting their résumé to websites like LinkedIn.com, Monster.com, Indeed.com, and even USAJobs.com (for federal positions).

When use of blockchains become commonplace, jobseekers will start posting résumés on their blockchains, but there will be one major difference—verification.

Validation and verification is one of the most valuable components of blockchains. Every industry requires some form of information verification. The question is, "what are blockchains verifying?"

In the case of recruiting and job search, blockchains will verify the education and employment information on the résumé. Blockchain education and employment verification will become more effective when all institutions that can verify information on a résumé or job application interface with blockchains—including educational institutions and employers. This is especially important for blockchain-related positions.

For example, according to *TechCrunch.com*, in April 2018, there were fourteen job openings for every one blockchain developer.[178] Organizations are coordinating blockchain career fairs where employers are looking for programmers as well as marketing, business, and legal professionals. Blockchain career fairs may be a part of a blockchain conference enabling attendees to expand their blockchain knowledge and look for blockchain positions in the same conference.

BLOCKCHAIN AND PAYROLL SYSTEMS SOLUTIONS

Employers have started to use blockchains to eliminate intermediary and payroll financial institutions. I even know employees who have been paid in bitcoin since 2015. Removal of intermediaries reduces costs, increases efficiencies, and streamlines business processes. In the future, blockchain payroll systems will streamline or even replace existing payroll systems.

Companies that use blockchains will not have to use banks or intermediaries to process their payroll because payroll can automatically run on blockchains. Blockchain payroll systems will greatly reduce the international banking costs for international companies that process payroll with multiple banks, and multiple banking fees, in many different countries.

If a blockchain payroll system seems far-fetched, this is not the first time businesses managed a major shift in their payroll systems. In the last two decades, businesses evolved from issuing payroll checks to using direct deposit. The next major evolution could be for businesses to evolve from direct deposit into banks to direct deposit into blockchain wallets. Payroll will automatically be directly deposited into the employee's blockchain wallet, which will give employees an excellent incentive to create an account in the blockchain.

In fact, the movement to pay employees started ten years ago in Kenya with a digital currency called M-Pesa. ("Pesa" means money in Swahili). Although the use of digital currency is very new to many parts of the world including the United States, Canada, and Europe, many countries in Africa, South America, and Asia have been using digital currency since 2007.

— ∽ —

In my travels to Africa, I had the privilege of talking with M-Pesa users about the positive impact M-Pesa has had on their everyday lives. In 2017, I had a memorable discussion with a Kenyan MBA graduate. During our discussion, he informed me many Kenyans, especially in rural areas, only earn a few dollars each day and if they are fortunate to have a bank account, they may have to travel long distances, sometimes more than an hour, to banks or other money transfer centers to access their money. Many rural families do not have access to banks because they do not meet Kenyan banking regulation identification requirements, or do not have enough money to maintain a minimum balance. With M-Pesa, workers in cities are paid in M-Pesa and immediately use their cell phone to send M-Pesa to their families in rural areas. When I asked the Kenyan MBA how long M-Pesa has been in use in Kenya, he responded, "ten years." I was honestly surprised because bitcoin also started ten years ago in 2008.

M-Pesa was created by Vodafone Safaricom, a Kenyan Mobile operator company, to resolve banking issues by using cell phones to text payments between users. Individuals can use M-Pesa to: receive wages, transfer money internationally, buy goods and services, obtain a loan, start a business, and make any other financial transactions using the M-Pesa platform.[179]

According to *CNN.com*, "M-Pesa: Kenya's Mobile Money Success Story Turns 10," in 2016, there were 30 million users in ten countries who processed six billion transactions at a peak rate of 529 per second.[180]

M-Pesa's implementation as a digital currency was ahead of its time. I had never heard of M-Pesa so during our conversation I tried but was unable to find it on CoinMarketCap.com. At first

I was surprised, but after a moment of thought, M-Pesa's role as a digital currency rather than an investment made sense. Since the vast majority of cryptocurrencies on CoinMarketCap.com dramatically increase and decrease in value, they are treated more like investments than currencies. As these cryptocurrencies are used for investment, they are not being used as currencies. Since M-Pesa is not used as an investment, it is more useful as a currency, which is good for the Kenyan people and economy. In another conversation with a Kenyan restaurant owner outside of Nairobi, I asked if she accepted M-Pesa? Her response, "M-Pesa is life. My business could not survive if I didn't use M-Pesa."

M-Pesa has expanded well beyond the borders of Kenya. According to Vodafone.com, M-Pesa is used in the following countries: Kenya, India, Lesotho, Mozambique, Romania, Tanzania, Albania, Democratic Republic of Congo, Egypt, and Ghana. This makes M-Pesa more than a local currency; M-Pesa is a regional currency.[181]

———

Blockchain payroll systems will vastly increase the use of cryptocurrencies. Employees will be required to create accounts in blockchains and start using the cryptocurrencies and blockchains for basic essentials. Then vendors, stores, and other commercial establishments will adopt blockchain technology to attract customers spending cryptocurrencies and tokens.

Blockchain technology's impact on business is very clear. When Buterin created Ethereum, he changed the landscape of business transactions, and one of the unintended consequences is blockchain technology's ability to level the playing field in business and employment, including diversity and inclusion.

DIVERSITY AND INCLUSION SOLUTIONS

As a labor and employment attorney and a Certified Diversity Executive, I provided corporate executives with many recommendations to increase diversity hires. During the application and hiring process, I advise employers to assign an employee to collect and sanitize the résumés for all gender, race, age, and other personally identifying factors and distribute the résumés to the hiring manager for the interview process. Most companies decide against using this approach as sanitization of résumés is very labor and time intensive. Enter the blockchain.

As the number of job postings increase on blockchains, blockchain technology's open nature, transparency, and verifiable structure will almost assuredly increase diversity in these organizations. Once a position is posted on a blockchain, the applicant can apply for the position and the public key will be the only identifying information. The interviewing company will not have any diverse information unless the applicant enters it on their blockchain. For example, including the names of their professional memberships may give insight to their gender, ethnic diversity, religion, or sexual orientation.

From a broader perspective, DAOs can increase diversity and inclusion as well as supply chain diversity. Currently, organizations are run by people and decisions are made based on communications and opinions. Centralized figures, such as CEOs, Executive Directors, or Members of a Board of Directors make the decisions for the organization.

In the DAO, smart contracts and the consensus protocols make decisions for the organization. Decentralized and autonomous decisions create an environment where all forms of diversity can thrive since personal opinions and unconscious biases are no longer part of the decision process. This is why I also call the

"DAO" a Diversity Advancing Organization. If DAOs continue to unfold as they have in recent years, I predict DAOs will have an impact on diversity and inclusion not because they actively advance diversity and inclusion, but because they give diversity a chance to thrive in a positive environment. Blockchain companies are leveraging the power of diversity in blockchain technology.

Gartner, a research firm, estimates blockchain technology is expected to add over $3.1 billion in business value, but the value proposition could involve more than money. STEAMRole, for example, uses blockchain technology and its own cryptocurrency, RoleCoin, for two purposes: providing STEAM-expert role models and a Diverse Talent Pipeline Platform (DTPP) for companies to use in tracking and hiring diverse talent.[182] STEAMRole connects role models with aspirants (called Steamers) particularly from groups that are underrepresented in STEM and the Arts (the "A" in the acronym). Both Steamers and role models are awarded RoleCoins for their activities, which incentivize participation and achievement. STEAMRole may not raise tens of millions in an ICO or have a high level of market share on CoinMarketCap, but STEAMRole is using blockchain technology to play in important role in professional development within the diversity community.[183]

CRYPTO TOKENS AND DIVERSITY COMMUNITIES

Diverse communities have started to use digital tokens to maximize or even increase their community purchasing power.

The LGBT Foundation is a Hong Kong-based non-profit that globally supports gay, lesbian, bisexual, and transgender individuals. The LGBT Foundation launched the "Pink Dollar," a blockchain utility token to generate capital to support LGBT individuals suffering from persecution around the world. The Pink Dollar will be used to support the following programs:

LGBT Identity Management enables LGBT members to verify and protect their identities, especially in countries where they face repercussions.

Pink Economy harnesses the economic might of the worldwide LGBT community by implementing a viable payment system for goods and services.

LGBT Impact sets aside a portion of tokens every year to generate funding to tackle LGBT discrimination, oppression, and inequality on a global scale.[184]

$Guap focuses on community economics and building a strong foundation for long-term economic wealth particularly for underserved communities like the black and brown communities. $Guap, urban vernacular for an abundance of money, is a utility token focused on building black businesses using smart contracts to incentivize performance of various tasks such as issuing "rewards."

The intent of $Guap, is to fund new businesses, historically black colleges and universities (HBCUs), and philanthropic efforts that use $Guap. $Guap users, including merchants, pass on savings to consumers, and benefit from gaining access to loyal customers.[185] According to Guap.com, $Guap focuses on four areas:

- The $Guap Market—A big data platform that collects and analyzes transaction information from blockchains so consumers in the black community view growth of community businesses as well as analyze risks.

- The $Guap Wallet—A multi-currency wallet focusing on coins such as $Guap and crypto education.
- Kowrii—A decentralized exchange that provides a simpler trading interface and educates "newbies" on crypto and crypto-trading.
- $GuapX—A music platform that empowers artists with the tools to distribute their music and monetize a fan base.[186]

Uulala, a minority-owned, US-based, social impact organization facilitates and accelerates the financial inclusion of underbanked and unbanked populations across the Americas. Uulala technology merges the 55 million U.S. Latino population that has a $2.4 trillion GDP with the 140 million Mexican population with a $1.4 trillion GDP. The Uulala platform empowers the underbanked communities of the world by providing financial tools to build credit, send money, participate in ecommerce, and rise out of a cash-only environment that lacks trust in traditional banking systems.[187]

CRYPTO TOKENS AND RELIGIOUS COMMUNITIES

Churches and religious communities are also leveraging blockchain technology to increase their fundraising efforts. In an interview with *Baptist News*, Christ Coin co-founder Emmanuel Ogunjumo shared his evangelical and humanitarian vision of Christ Coin. Christ Coin is not affiliated with any denomination. Ogunjumo states: "We launched [Christ Coin] to be independent of any particular denomination because we want everyone to benefit from Christ Coin."[188]

Christ Coin was designed to serve as a church currency and a global currency. Christ Coin is also an example of the number of blockchain ventures that are not able to sustain long-term business success. When I first started writing *Blockchain or Die* in 2017, Christ Coin was a hot topic in the religious and blockchain community. During the editing phase of *Blockchain or Die*, I revisited all references, including Christ Coin, only to find that the Christ Coin websites were gone and I could not find any recent social media posts. I do not know what happened to Christ Coin, but it is safe to say at the time of publication Christ Coin is not currently active.

Christ Coin, and other religious based cryptocurrencies, opened the world of cryptocurrencies and blockchain technology to segments of populations that may not otherwise have exposure. In some communities, the church is the main source of education and social interaction and its importance in educating its congregation on cryptocurrencies should not be underestimated. In these communities, the church is the perfect organization to expose blockchain technology to congregations and communities.

At the risk of combining church and state, the next section discusses legislation and blockchain technology.

LEGISLATION, BUSINESS, AND BLOCKCHAIN TECHNOLOGY

Similar to corporations, industry leaders, and diverse businesses, legislators and policy makers are paying attention to blockchain technology. Legislation can be the best friend or worst enemy to business and industry, especially when it relates to new technologies. For example, at the time of publication, the following countries have ruled bitcoin is illegal: Afghanistan, Algeria, Bangladesh Bolivia, Pakistan, Qatar, The Republic of Macedonia, Saudi Arabia, Vanuatu, and Vietnam. In these countries, it is impossible for individuals or businesses to use cryptocurrencies.[189] These examples show the negative impact governments can have on cryptocurrencies and blockchain technology.

As more businesses use blockchain technology, the faster it will become a fully integrated and functional business solution. If legislators pass laws mandating use of blockchain technology to collect, store, and distribute information, blockchain use will dramatically increase.

For example, Chapter 1 provided a comparison of the blockchain process to the real estate title search, and blockchain technology will only work in real estate titles if every company is required to file a lien on a blockchain against a homeowners' title. Otherwise, the title search in that blockchain would be incomplete as liens may exist outside of the provided blockchain. The best approach is for state and federal legislators to pass legislation mandating the filing of all liens on blockchain technology. In addition, local and statewide property records offices must transfer all existing liens to a blockchain. As long as businesses have the option to file a lien outside of the provided blockchain, a blockchain based title search will be ineffective.

LEGISLATION PAVES THE WAY FOR BLOCKCHAIN TECHNOLOGY

Blockchain legislation will solidify the use of blockchain in the public and private sector and advance the use of blockchain at a faster rate. Forward thinking legislative bodies have already passed blockchain related legislation. For example, Wyoming passed five blockchain related bills in 2018.[190] State legislatures are covering many blockchain categories including, but not limited to: taxes, voting, corporate filings, money transmitters, and legalities of blockchain contracts. The following section will focus on two types of state legislation coordinated by year and state: legislative studies/working groups and corporate regulations.

LEGISLATIVE STUDIES AND WORKING GROUPS

By passing laws to conduct studies, create working groups, or revise corporate regulations, legislators will understand the impact of blockchain related legislation on their governance models and constituents.

The following table in figure 20 shows legislation passed in the state legislatures for 2018.[191] You should note that legislative action was taken as early as 2015 by the state of Vermont, but didn't really start to gain momentum until 2017 with additional legislation in Delaware, Hawaii, Illinois, Maine, New Jersey, New York, Virginia, and Wyoming.

STATE	LEGISLATION	SUMMARY
Connecticut	S.B. 443 Signed by governor 6/6/18, Special Act 18-8	Establishes the Connecticut blockchain working group; develops a master plan for fostering the expansion and growth of the blockchain industry in the state; recommends policies and state investments to make Connecticut the world leader in blockchain technology.
Delaware	S.B. 182 Signed by governor 7/23/18	Updates the Delaware Revised Uniform Limited Partnership Act to provide specific statutory authority for Delaware limited partnerships to use networks of electronic databases (examples of which are described currently as "dis-

STATE	LEGISLATION	SUMMARY
		tributed ledgers" or a "blockchain") for the creation and maintenance of limited partnership records and for certain "electronic transmissions."
Delaware	S.B. 183 Signed by governor 7/23/18	Updates the Delaware Limited Liability Company Act to provide specific statutory authority for domestic limited liability companies to use networks of electronic databases (examples of which are described currently as "distributed ledgers" or a "blockchain") for the creation and maintenance of limited liability company records and for certain "electronic transmissions."
New Jersey	A.B. 3613 S.B. 2297 Passed Senate 12/17/18	Establishes NJ Blockchain Initiative Task Force.
New York	A.B. 8792	Directs the state board of elections to study and evaluate the use of blockchain technology to protect voter records and election results.

STATE	LEGISLATION	SUMMARY
New York	A.B. 8793	Relates to establishing a task force to study and report on the potential implementation of blockchain technology in state record keeping, information storage, and service delivery.
Vermont	H.B. 765	Proposes to implement strategies relating to blockchain, cryptocurrency, and financial technology in order to: promote regulatory efficiency; enable business organizational and governance structures that may expand opportunities in financial technology; and promote education and adoption of financial technology in the public and private sectors.
Vermont	S.B. 269 Signed by governor 5/30/18, Act 205	Modifies the definition of "blockchain" and "blockchain technology"; enables the creation and regulation of personal information protection companies; creates studies for expanding the use and promotion of blockchain technology; enables the creation of blockchain-based limited liability compa-

STATE	LEGISLATION	SUMMARY
		nies; and creates a study for the potential use of blockchain technology in government records.
Virginia	H.J.R. 153	Establishes a one-year joint subcommittee consisting of seven legislative members and five non-legislative members to study the potential implementation of blockchain technology in state recordkeeping, information storage, and service delivery. In conducting the study, the joint subcommittee shall research, analyze, and consider (i) opportunities and risks associated with using blockchain technology in state recordkeeping, information storage, and service delivery; (ii) different types of blockchain technology and the feasibility of implementing each type; (iii) projects and use cases currently under development in other states and nations and how those cases could be applied in Virginia; (iv) how early adoption of blockchain tech-

STATE	LEGISLATION	SUMMARY
		nology may stimulate interest and growth in Virginia's information technology industry; and (v) how current laws in the Commonwealth can be modified to support blockchain technology.
Wyoming	H.B. 1 Signed by governor 3/14/18 with line item veto 3/14/18, Chapter 134	Creates a blockchain task force to identify governance issues related to blockchain technology and develop appropriate legislation to be recommended to one or more appropriate legislative committees for consideration.

Figure 20 Blockchain related legislation introduced in 2018

In 2019, at the time of this writing, twenty-seven states introduced seventy-two laws, of which fifty-nine were passed, signed into law by the governor, or are still pending. New York is the most active state with five active blockchain related bills.

ADDITIONAL BLOCKCHAIN RELATED STATE PROGRAMS

Some states are doing more than creating working groups to incorporate blockchain technology into their state operations. In November 2018, Ohio was the first state to accept bitcoin and other cryptocurrencies as a means for businesses to pay their state taxes. According to the *Forbes* article, "Ohio Becomes the First State to Allow Taxpayers to Pay Tax Bills Using Cryptocurrency," Ohio businesses can use OhioCrypto.com to pay business taxes. Ohio used Bitpay to process the cryptocurrency payments and Bitpay converts them to U.S. currency that is deposited in the State of Ohio's account. Ohio's general use of cryptocurrencies is limited as Ohio is not holding, mining, or investing in cryptocurrencies.[192] This is a major step for a state government—Ohio moved past using working groups to research the use of cryptocurrencies to implementing a statewide tax payment program.

West Virginia is another state that successfully implemented a blockchain program. Govtech.com reports in November 2018, West Virginia became the first U.S. state to use blockchain technology in a federal general election. The blockchain-voting pilot was limited to military and other voters living overseas. One hundred forty-four West Virginia voters stationed in thirty countries voted in the 2018 midterm election.[193] As blockchain technology matures, many states will report the successful completion of blockchain programs.

These additional state programs are important first steps as they are two areas that effect everyone's life: taxes and voting. If blockchain technology has the power to impact these areas, it has the power to impact every area of our lives.

SUMMARY

The vast majority of the public has yet to realize the major role blockchain technology will play in their lives, but as more companies roll out blockchain products, the benefit of blockchains will become obvious. This chapter showed the real life applications of blockchain technology in health care, cloud storage, and human resources (three areas that directly impact most of our lives), but the Wal-Mart supply chain example shows how blockchain technology will impact food from farm to table.

Diverse businesses are serving as blockchain resources and advocates in their communities. This book only covered LGBTQ, African-American, Latino, and religious communities, but many other diverse communities are using blockchain technology to empower the social and financial ties in their community.

Finally, legislation will play a critical role in the future of blockchain technology as they decide the best way to regulate this new technology. Legislators that choose to partner with businesses to understand the complexities, benefits, and risks of blockchain technology will draft legislation that will enable blockchains to accomplish great things is every industry.

SUCCESS STRATEGY ACTION ITEMS

1. Visit the Hyperledger website: www.hyperledger.org. Hyperledger is an excellent source for exposure to the blockchain and an overall good way to use the information in this chapter.

2. Visit IBM.com/blockchain/solutions to learn more about IBMs blockchain solutions. IBM's innovative ideas may give visitors ideas for blockchain use in their organization.

3. Research and read medical blogs and articles. Also consider researching whether your medical providers or insurance companies are considering the use of blockchain technology.

4. As discussed in the cloud storage section, you don't have to invest money into cryptocurrencies to profit from cryptocurrencies; you can use your computer to provide storage for a cloud storage blockchain company.

5. Research and read human resources blockchain blogs and articles. Many articles will describe blockchain technology's impact on different areas of human resources.

6. Conduct an internet search for blockchain positions. The search results will update you on the market for blockchain experts. Human Resources departments are actively posting positions for people with experience in blockchain technology.

7. Research blockchain companies in your specific area of diversity. The three companies referenced in this chapter represent a small of portion of diverse companies using blockchains as their primary business model to benefit the diversity community.

8. Follow the progress of blockchain legislation. There will be many state, national, and international laws related to blockchain technology. Any research into blockchain technology must include up-to-date information and understanding of blockchain legislation. This is a crucial timeframe for the use and maturity of blockchain technology as legislatures are playing an important role in the scope of use related to it.

[7]

THE BLOCKCHAIN EVOLVED

"Design isn't finished until someone is using it."[194]

—Brenda Laurel, PhD, Independent Scholar

BLOCKCHAIN TECHNOLOGY IS MORE than a series of algorithms, consensus protocols, and distributed ledgers to manage transactions; blockchain technology is a powerful tool for change, but it is useless unless people are using it. The average person is not going to use a blockchain to start a company, invest in ICOs, or open a wallet to invest in cryptocurrencies. The average person will use blockchains to make everyday life a little bit easier, secure, or even more fun. This experience is user interface, and this is how blockchain technology is evolving.

WHAT IS USER INTERFACE?

In everyday terms, user interface is when the user interacts with the technology and software. Smart phones and applications increased user interface by giving users easier and faster access to the internet. Blockchains are also improving user interface.

For example, Brave, a blockchain-powered web browser, bridged the blockchain user interface gap by creating a web browser that blocks trackers to prevent the unauthorized collection of personal information and blocks ads that use data and processing power. Brave also uses the Ethereum based token, Basic Attention Token (BAT), which can be used between advertisers, publishers, and users. According to BasicAttentionToken.org, the BAT can be used to "obtain a variety of advertising and attention-based services on the BAT platform. The utility token is based on user attention, which simply means a person's focused mental engagement."[195]

Initially, blockchain user interface was very complicated. In the early years, mostly coders and programmers were using blockchain technology because the technical community created bitcoin and blockchains. Blockchains became a success because, compared to bitcoin, blockchains presented solutions to major business challenges. Ten years later, the world is looking for blockchain-enabled programs with high user interface. CryptoKitties was one of the first companies to improve blockchain user interface.

CRYPTOKITTIES, THE FIRST BLOCKCHAIN GAME

In 2010, cryptocurrencies and blockchains were primarily focused on the financial industry, so most users were experts in technology or finance. This is where blockchains started, but not where it needs to finish. The next step was to design non-financial blockchain based programs with improved user interface so the average person would have an interest and the ability to use a blockchain. CryptoKitties was one of the first blockchain applications to start the transition to an improved blockchain user interface.

I first learned of CryptoKitties a few years ago when they presented at the Government Blockchain Association monthly meeting. At that time, all of the cryptocurrencies were rapidly increasing in value so I was focused on cryptocurrencies as an investment and blockchain technology as a business tool, I was not thinking about funny looking kitties on a blockchain. But soon after I started trying to use blockchains, I realized the importance of user interface, and the importance of blockchain applications like CryptoKitties. It was time to pay CryptoKitties.co a visit.

The first thing CryptoKitties.co visitors see are colorful and creative digital cats sitting down and looking at you with big rounds eyes. But CryptoKitties are more than digital cats on a blockchain. CryptoKitties is a marketplace where digital cats are bought and sold and once purchased, owners can breed, sire, or sell their CryptoKitties.[196]

The CryptoKitties White Pa-Purr: *Collectible and Breedable Cats Empowered by Blockchain Technology* (hereinafter "The White Pa-Purr") states CryptoKitties will make blockchain technology accessible to the average consumer through four main tactics:

- Gamifying features that leverage blockchain's unique applications.
- An approachable, consumer-facing brand based on genuine a passion for blockchain technology.
- An open platform inclusive to users of all levels of technical knowledge.
- A sustainable revenue-based model (as opposed to an ICO).[197]

The White Pa-Purr goes on to state:

"By normalizing the practical application of smart contracts and cryptocurrency transactions, [CryptoKitties] will empower everyday consumers with a basic fluency in distributed ledger technology. Likewise, by showcasing a practical use for block-chain technology outside of the financial industry, we hope to broaden the public's understanding of the technology and its pubic application."[198]

The White Pa-Purr abstract points out the importance of a user-friendly blockchain. First, note The White Pa-Purr abstract is written in very plain language. Unlike the bitcoin and Ethereum abstracts, the CryptoKitties abstract does not use technical terms like smart contracts and nodes. It was important for the Bitcoin and Ethereum White Papers to create these new terms, but in making blockchain technology accessible to the average consumer, the authors of the CryptoKitties published a white paper to appeal to the average consumer. CryptoKitties expanded blockchain technology into the multi-billion dollar tech gaming industry.

CryptoKitties is taking the next steps to evolve with block-chain technology. CryptoKitties will be able to do more than just sit on your screen or phone, in the CryptoKitties DApp; they will be able to fight each other in matches for and kittiefight tokens.[199]

Finally, while other cryptocurrency and blockchain companies were raising funds through the ICO process, CryptoKitties used a revenue-based model to monetize their blockchain application. As a result, CryptoKitties is not a security.

BLOCKCHAINS—THE NEW ART DEALERS

If you are not into digital feline blockchain crypto collectibles, CryptoPunks offers a human version of crypto collectibles. CryptoPunks are 24x24 pixel art, mostly punky looking headshots, with the throwback look to the digital Atari days, circa 1980. Although most are human headshots, there are a few apes, zombies, and aliens in the CryptoPunks collection. All CryptoPunks are unique and stored on the Ethereum blockchain. The first 10,000 CryptoPunks were free, but they were quickly claimed and have to be purchased from the Ethereum marketplace.[200] Although CryptoPunks are digital art, they can also be used as an investment.

CryptoKitties and CryptoPunks have opened the door for digital art on the blockchain. Although the CryptoKitties and CryptoPunks crypto collectibles are not paintings or sculptures, they are works of art—specifically digital art.

Now that the blockchain digital art door has been opened, how does digital art evolve on a blockchain? Is there a place for traditional art on a blockchain? Can a blockchain provide intellectual property protection for works of art on a blockchain?

Chapter 1 discussed how bitcoin used blockchain technology to solve digital copying and the double spending problem. Digital copying does not only apply to digital currencies, it can include any digital product that has value, including art. The unauthorized copying of digital art threatens the two components that gives art value: exclusivity and verifiability. Experienced art connoisseurs and investors will not spend a great deal of money on art that is not exclusive or cannot be verified. But if everything placed into a digital format can be copied, how can art remain exclusive and verifiable? This is one of the reasons Christies is exploring the use of blockchain technology.

The decentralized, transparent, immutable components of the blockchain protect the exclusivity and verifiability of art on the blockchain. When one thinks of fine art and auctions, "Christies" is one of the first words that comes to mind. Founded in London in 1766, Christies' iconic brand is steeped in tradition, and recently, Christies added one more title to their iconic brand: blockchain innovator. In July 2018, Christies hosted a one day Summit, "Art + Tech Summit—'Is the Art World Ready for Consensus,'" which addressed the issues of the blockchain on the art world.[201] Since blockchain technology is transparent and immutable, art buyers can use blockchains to determine the authenticity of a piece of art, access the sale history, and tap into any records associated with the art. Blockchain technology can also reduce unauthorized copies of artwork that will later be sold to unsuspecting art buyers. In a system where art buyers have to rely on intermediaries to certify the authenticity of a piece of valuable art, blockchains will allow buyers to self-certify potential art purchases.

Every other company in the world should be paying attention to Christies' interest in blockchains. If a 252 year-old auction house is looking for ways to incorporate blockchain technology, every company can engage in the same exercise.

BLOCKCHAINS ARE ALIVE WITH THE SOUND OF MUSIC

Visual art is not the only art on a blockchain; music is on a blockchain as well. Music on a blockchain is the latest in the love-hate relationship between digital technology and the music industry. Napster's peer-to-peer sharing protocols forever changed the music industry and when the music industry saw a decrease in sales and profits, they had to evolve. The music industry found profits in online sales, ring tones, and streaming platforms like

Apple Music, Pandora, and Spotify, but with blockchain technology, the music industry had to evolve again.

The music industry is full of intermediaries that have very little to do with creating the music and everything to do with making the music accessible the buying public. Essentially, the music industry is a collaboration of companies that create, distribute, and market music to the masses. Blockchain companies like Choon and Voise are creating new methods to increase access to music and eliminate the music industry business intermediaries.

According to Choon.co, "Choon is a music streaming service and digital payments ecosystem—designed to solve the music industry's most fundamental problems." Choon runs on the Ethereum platform and uses crypto tokens called "notes" for Choon subscribers to listen to music on the Choon platform. Once a subscriber streams a song, Choon uses "smart record contracts" to pay the artists with notes into their wallets.

Choon also pays subscribers in notes to listen to sponsored songs. With 343 genre playlists, Choon has an impressive variety of music for almost every musical taste. In December 10, 2018, Choon had 10,630 artists earning notes, 34,523 tracks and artists have earned 83.6 million in notes.[202] As of July 21, 2019, Choon has 12,537 artists earning 168.3 million in notes from 46,639 tracks.[203] These figures clearly show Choon's blockchain presence is growing,

Voise.co, another blockchain music business, describes Voise as "a blockchain powered anonymous decentralized platform with personalized token based on Ethereum's smart contract ecosystem for transactions for the music industry that allows artists to monetize their work in a collaborative P2P marketplace. Artists set a price for their works, provide free sample tracks, and seek

support from music enthusiasts and users on the platform."[204]
The Voise process is very straightforward:

1. The artist uploads the content.
2. Users see relevant music.
3. The user pays with Voise tokens.
4. Artists earn nearly 100% of the revenue.[205]

Similar to other digital music platforms that primarily use apps to provide music, Voise has a DApp to increase user interface.

Unlike most of the companies in this chapter, Voise has a crypto token trading on CoinMarketCap.com.[206]

There are other DApps for recorded music on a blockchain including: Cryptotunes, Lava, Music Coin, and UJO Music. Research the DApp that best suits the way you prefer to listen to music.

BLOCKCHAINS ON THE CATWALK?

Most of the fashion industry seems like glitz and glamor but when the glitz and glamor are put aside, manufacturing, supply chain, and retail are the real nuts and bolts of the fashion industry. Sounds like a perfect fit for blockchain technology.

Chapter 6 discussed Wal-Mart's blockchain actions to track food shipments and improve food safety; the same supply chain and inventory control principles apply to the fashion industry. Similar to the art and music industry, the fashion industry suffers from counterfeit products, but there are also important differences. The products for the fashion industry are based on fabrics and other materials while the music industry is primarily digital. Compared to the art industry, most fashion industry profits are based on selling large quantities of merchandise while the high-end art industry places the profit premium on original works of art and limited editions.

For example, Provenance's fashion blockchain solution allows businesses and consumers to track physical products through labeling and smart tags. When a consumer uses their smart phone to scan a Provenance smart tag on an item of clothing, the smart tag will tell the consumer the clothing's entire trip to the point where it ends up in the store. The transparent, verifiable, immutable nature of blockchains and the smart tag will prevent companies from selling counterfeit merchandise. Once the consumer scans the merchandise, they will be able to verify its authenticity.[207]

PLACE YOUR BETS ON BLOCKCHAINS

Gambling is another industry that has adopted the use of blockchain technology. The gambling industry is big money, and gambling on a blockchain means big money as well. Once a blockchain was created and commercialized, gambling companies started to structure gambling on blockchains. Two blockchain based gambling sites include Bet Dice and Endless Dice.

Bitcoin casinos are big business as well, but they are not for every bitcoin user. Gamblers that prefer to gamble in fiat currencies will not find blockchain casinos a viable alternative. Using the blockchain basics we have discussed so far, the pros and cons of cryptocurrency and blockchain casinos follow:

Pros:
- No third parties or intermediaries involved in transactions
- All transactions and cryptographically secured
- Players can use cryptocurrencies for anonymous play

Cons:

- According to the IRS, cryptocurrency winnings are "property" so winners have to pay taxes.
- At this time, cryptocurrency payments are irreversible.

Finally, remember that cryptocurrency prices are highly volatile. This is neither a pro nor a con, but a cryptocurrency that you may use to gamble, may also appreciate in value the next day.

Similar to CoinMarketCap and State of the DApps serving as portals to cryptocurrency and blockchain companies, Casinoblockchain.io serves as portal to eight different online casinos and sports betting of which four are blockchain based:[208]

Casino Fair
Games: Slots, Roulette, Poker, and Blackjack
Currencies: FUN, ETH
License: Curacao

One Hash
Games: Sports Betting, Dice, and Slots
Currencies: BTC
License: Curacao

Edgeless.io
Games: Blackjack, Dice, Slots, and Baccarat
Currencies: FUN, ETH
License: Curacao

True Flip
Games: Lottery
Currencies: BTC, FUN, ETH
License: Curacao

This list includes the major components of blockchain gambling: games, currencies, and location of licenses, and this list will only increase globally as more blockchain companies and casinos adopt blockchain technology.

DATING ON A BLOCKCHAIN

The use of online dating and dating apps has increased over the years and so has fraud and abuse. Examples include:

- Catfishing: Deceptive online romances by creating a false profile, and even a false gender.
- False age declarations: Declaring they are older to attract an older dating pool, or younger to attract a younger dating pool.
- Money scams: Creating profiles to collect phone numbers and emails to send requests for money.

These, and many other examples of fraudulent dating activities, show the need for transparency and verification of users to ensure a safe, fraudulent-free, online dating experience. As discussed in previous chapters, anytime a centralized system stores credit card or personal information, hackers will attempt to illegally obtain it. Unfortunately, this includes online dating sites. The online dating industry is another industry that will benefit from blockchain technology.

Blockchain technology's verifiable, transparent, and immutable components are a "perfect match" for online dating. The verifiable nature of blockchain technology means every user's account must be independently verified on the blockchain. If a user tries to create one or more fraudulent accounts, the account will not be verified by the users.

Blockchain technology's transparent nature means the other blockchain daters will share the users information. Blockchain technology's immutable nature will prevent users from fraudulently changing their information once the information has been verified.

Finally, hackers will have very little interest in a blockchain dating app as the user will not have any personal information on the app. Viola, Love Block and Ponder created solutions to bring love to blockchains.

Viola.ai is more than a typical singles dating site because users can include singles, unmarried couples, and married couples. Viola also has an app listed on StateoftheDApps.com. How is Viola.ai used beyond the scope of the singles scene? Viola.ai states their technology can be used for the following:

- Singles—Access curated & verified matches
- Unmarried Couples—Attached couples can now marry on the blockchain; and use smart contracts for their union
- Married Couples—Declare your love and commitment on the blockchain with immutable Smart Marriage Contracts[209]

So far Viola.ai's footprint is in Asia, specifically: Singapore, Malaysia, Hong Kong, Thailand, Indonesia, and Japan. Viola.ai's goal, however, is to achieve a global presence with 25+ million users by January 2020.[210]

Ponder is another blockchain based dating company. Ponder uses matchmakers and blockchain technology to fix people up. Matchmakers receive financial rewards ($10 if the couple likes each other) and ($1,000 if they end up getting married).[211]

BLOCKCHAINS AND BOOTY CALLS

When I first learned about blockchain technology, I never thought I would see the words "blockchain technology" and "sex" in the same sentence, but here it is as a part of my book. Since blockchain technology is going to change every aspect of our lives and impact every industry, the sex industry, is not an exception.

How does a blockchain impact sex? Essentially, sex is a physical, and perhaps an emotional, transaction between two or more people. Whether sex is for love, money, passion, or any other reason I cannot think of, sex is a contract. Contracts require a meeting of the minds and sex contracts require meeting of the minds and bodies.

In the initial stages of writing *Blockchain or Die* I found LegalFling, a blockchain company that created binding agreements for parties before they engaged in sex. However, when I went back to fact check LegalFling during the editing phase, the website and app was no longer available. LegalFling.io posted a message that it was removed from Google Play and the iStore. LegalFling may resurface under a different application system, but at this point, LegalFling is not in use.

Spank Chain, an adult entertainment blockchain company, uses blockchain technology to provide a safer, more secure adult entertainment experience. Adult entertainment is a leading subject for traffic on the internet, and like it or not, the adult entertainment industry successfully implements the user interface. Spank Chain's mission is to "deploy a framework which provides the core benefits of blockchain technologies—privacy, security, self-sovereign identity, and economic efficiency—to the adult entertainment industry."[212] For example, Spank Chain is creating a game called "Proof of Spank" to address under age model swapping and incentivize SpankChain usage by providing the security that all users are eighteen years or older.[213]

Exotic dancers are the final topic in area of blockchain and sex. Exotic dancer patrons no longer have to use dollar bills to tip dancers, now they can use cryptocurrencies. The site Bitcoin-ist.com reports that patrons can scan QR code tattoos on the dancers for tipping in cryptocurrencies. This seems to be a win-win as patrons prefer discreet transactions and dancers that earn large sums of money do not have to explain large cash deposits to banks.[214]

YET ANOTHER USE FOR SMART PHONES

Smart phones were a major technology game changer. Smart phones combined many different technologies (cell phones, computers, video games, cameras, etc.) into one device giving millions of people who could not afford a computer access to the internet. Millions of people were able to shop online, talk to friends and family across the world, record high quality videos, and transfer money. As smart phones evolved, they increased internet access and user interface. Cellphone manufacturers realized they held one of the keys to increased user interface and started to develop blockchain-enabled smart phones.

Two companies, HTC Corp and Sirin Labs, released block-chain-enabled smart phones to serve as a wallet to store cryptocurrencies and act as a node on a blockchain. The name of the HTC phone is the "Exodus" and HTC wants the Exodus to "double or triple the number of nodes of Ethereum and bitcoin."[215] The Sirin Labs phone is called the "Finney," named after cryptocurrency pioneer Hal Finney, and has the same function to store cryptocurrencies as a cold-storage wallet.[216] Also, both phones allow users to play DApp games, such as CryptoKitties. Blockchain-enabled smart phones will greatly increase user interface. Once thousands of blockchain-enabled phone users increases to millions then tens of millions, blockchain user interface will occur many times over every second of every day.

But as of the time of this writing, blockchain smart phones have not been widely accepted and have limitations. For example, shop.sirinlabs.com includes the disclaimer: "TCS requires KYC and has currently been blocked under local government regulation for some countries. We are working with local government regulators to enable it." Functionality on Verizon network is currently limited (text messaging).[217]

SUMMARY

A few years ago I was in my best friend Rob's car and his three-year old son, Carter, was in the back car seat. While Rob and I were talking, Carter said, "Phone Daddy, phone." At a stoplight Rob gave me his iPhone and told me to give it to Carter.

I thought I was condemning the iPhone to a death by toddler incident, so I asked, "Are you sure you want to give Carter your phone?"

Rob replied, "Yeah, he knows how to use it."

Still doubtful, I passed the iPhone to Carter. Carter took the phone, unlocked the screen, and started swiping left until he got to the Angry Birds game. Then Carter opened the game and started to play. Briefly, Carter's gaze caught mine as his eyes innocently said, *That's what you get for doubting my abilities.*

This story is the ultimate example of user interface. When a three-year old child is able to access a blockchain, use DApps, or any other blockchain products, blockchains would have achieved ultimate user interface. But all of my blockchain learning experiences were not as innocent as Carter and Angry Birds.

While conducting research for this book I found an article that reported a major airline was using Ethereum to replace their frequent flyer program. This article caught my attention as I am regularly on a plane crossing the globe and the thought of accruing ether for my travels was very attractive.[218] When I visited the major airline site to verify the blockchain-based frequent flyer program, it was non-existent. Confused, I went back to the article to review the citation and sources, and then I saw "Happy April Fools" at the bottom of the article. I was half amused and half disappointed. This experience has at least two lessons for anyone that reads this book.

1. Always double check and triple check resources. I usually check the company website to confirm reports on their cryptocurrency or blockchain activities.

2. People are always eager to learn new ways to use new technology. I was eager to see how the major airline was going to use ether as the currency for their frequent flyer program. Truth be told, while this article was an April Fools joke, I was convinced there was a business case to use blockchains to evolve frequent flyer programs as frequent flyer miles are a form of digital currency.

A few months later, I discovered Singapore Airlines partnered with KPMG to launch KrisPay, their own blockchain wallet. *TechinAsia.com* reports in 2017, KrisFlyer members accumulated $700 million USD worth of air miles. By comparison, KrisFlyer members accumulated $573 million in 2014 and $445 in 2011. But KrisPay is much more than a blockchain-based frequent flyer program, because KrisPay can be used with local retailers and can even be converted to cash. Looks like the April Fool is no longer a joke, but maybe it was a bit ahead of it's time as Singapore Airlines later temporarily suspended the partnership between KrisPay, the airline, and the app. I am of the opinion that partnerships between companies that support blockchain-based frequent flyer programs and airlines will benefit all parties involved, especially the frequent flyers; it's only a matter of time before it takes off.

The companies in this chapter are the future of blockchains and the future of business. Many of the companies are in the early stages of development so they have many hurdles before they move from design to actual use, but they are on the right path.

SUCCESS STRATEGY ACTION ITEMS

1. Download and use Brave as your internet browser. Brave will serve the same function as Google Chrome, Safari, Firefox, or any other internet browser; however, Brave will not track your internet locations and you will no longer receive pop up ads.

2. Visit CryptoKitties.co and CyberPunks.ai. If your only exposure to blockchain technology has been cryptocurrency trading platforms, these blockchain sites will show a different side of blockchains. Both websites combine blockchain usage, digital art, and financial incentives to show people how to use blockchain technology.

3. If you are a music fan, blockchain music technology is the latest way you can access music online. As the number of blockchain music providers increases and user interface improves, you will be able to steam music as easily as you do with iTunes or Pandora.

4. Similar to the internet, there may be many blockchain sites that you would normally visit. Some of those blockchain websites include: gambling, dating, and sex-based sites. While you do not have to actually use these sites, you can visit them to experience the user interface.

5. Download and listen to music and videos from a blockchain.

6. If you are interested in blockchain-enabled smart phones, use the information in the chapter to research your options.

[8]

INCORPORATING BLOCKCHAIN TECHNOLOGY INTO YOUR BUSINESS

"We need to come up with use cases for this technology that drive clear benefits for individuals and institutions—these are our customers. Too often we see bitcoin and blockchain technologies as solutions in search of a problem. We don't just need these systems to be technically better than the alternatives—we need them to be more user-friendly."[219]

—Abigail Johnson, Business Woman

THIS IS AN EXCITING TIME for businesses. The incorporation of blockchain technology into the current business structure is the daily hot topic, but the reality remains that blockchain technology is still in its infancy. In the nineties, the internet dramatically changed the way companies conducted business; blockchain technology will have a very similar, if not greater impact. However, businesses must incorporate blockchain technology for the right reasons and in the right way. Abigail's quote is essential because it puts the incorporation of a blockchain into perspective. Blockchain technology should not be a solution in search of a problem; it should be solution to a problem.

What Role Should Blockchain Technology Play in Your Business?

Your business will not benefit from blockchain technology unless it is user friendly and solves actual problems. Of course, most new technologies, including blockchain technology, come with a steep learning curve and ramp up in technology, so it takes time to become user friendly.

It's simply too early in its development for the vast majority of non-blockchain businesses to fully adopt and incorporate blockchain technology, but it's never too early to research and plan to be ready for when the time is right. The *Forbes*.com article, "Is Blockchain the Invisible Answer to Your Blockchain Needs?" quotes an IBM survey that "indicates about one-third of executives are either currently using blockchain or considering incorporating it into their companies. Of the 3,000 executives who took part in the survey, 80% planned to use a blockchain solution in response to financial shifts or to develop new business models."[220]

These survey results are a clear indication that corporate executives are incorporating blockchain technology into their current or future business strategy. In the near future, these corporate decision makers will be searching for blockchain programmers, consultants, trainers, and attorneys. Will your business be ready to supply the new blockchain demand? How should you prepare your business to use blockchain technology? This chapter addresses these forward-thinking questions.

BUSINESSES BUILT ON A BLOCKCHAIN

Although blockchain technology is relatively new, some entrepreneurs have already established successful blockchain-based businesses. For example, over the past few years, ConsenSys established itself as a leader in blockchain consulting and programing. ConsenSys uses decentralized technology, specifically Ethereum, to build and scale tools, disruptive startups, and enterprise software products. ConsenSys.net states their vision is "that blockchain technology allows us to progress to the next generation of the Web that we call 'Web 3.0'".[221]

One example of this mission in action is the Brooklyn Project, which ConsenSys describes as "an industry-wide initiative to promote token-powered economic growth and consumer protection."[222] Companies like ConsenSys and projects like the Brooklyn Project provide the innovation and leadership required to advance blockchain technology into an established business practice and technology. But the ConsenSys model is not for most businesses; so should your business incorporate blockchain technology? One of the first ways to determine this is to use a blockchain business model.

BLOCKCHAIN BUSINESS MODELS

There are a number of blockchain business models designed to answer the core question in this chapter: "Should I incorporate blockchain technology into my business?" Of course, the answer is: "it depends." The *Medium*.com article "When Do You Need Blockchain? Decision Models," describes ten commonly referenced blockchain models:

1. Birch Model
2. Birch-Brown-Parulava Model

3. Suichies Model

4. IBM Model

5. Lewis Model

6. Sebastien Meunier 2017 Model

7. Karl Wüstl and Arthur Gervais Model

8. Morgen E. Peck Model

9. DHS Model (Department of Homeland Security)

10. Cathy Mulligan Model[223]

The article encourages readers to review all of the blockchain models to assist in the blockchain decision process. This chapter focuses on three of the ten blockchain models: Birch Model, Suichies Model, and DHS Model. All three models present a very different blockchain decision process. The Birch Model is the most basic, the Suichies Model examines the best blockchain solution based on information, and the DHS Model provides the best solution when sensitive data is present.

BIRCH MODEL

While there is a humorous component to the Birch Model, it does make a valid point. Most companies do not need blockchain technology, at least not yet. Many start up businesses have consulted with me having the desire to add a blockchain component to their business. I essentially ask them "why do you need a blockchain to meet this business goal?" Many times they provide an answer, but it usually a business goal that can be met with current technology.

For example, a small startup contacted me on LinkedIn and after a few emails we set up a fifteen-minute phone consultation. What follows is a summary of the conversation.

"I want to use the blockchain to raise money for my business."

"What kind of business?" I asked.

"We provide business marketing and event planning," they replied.

"How much technology do you use in this business?"

"Not much," said the business owner, "a website, social media, and a cloud storage database."

"Have you tried to raise money any other ways? For example, crowdfunding?"

"No," the owner said.

"I recommend you consider other means to raise money," I began. "If you are looking to raise money through an ICO, you are looking a spending a lot of money on legal fees for SEC compliance as well as a robust marketing campaign."

"Thanks for your time."

By way of example, this company did not need a blockchain, but in ten to twenty years, every company may need it; similar to the way every company currently needs access to the internet and a webpage. When blockchain technology becomes an integral component of business transactions, the model may change from "When do I need blockchain?" to "Do I need a blockchain?" Then the answer will be "yes."[224]

SUICHIES MODEL

The Suichies Model (see figure 21) asks the question "Do you even need Blockchain?" In addition to answering this question, assuming the answers to the questions in the flowchart determine a blockchain is needed, the Suichies model provides the block-chain options referenced in Chapter 5 (public blockchain, hybrid blockchain, or private blockchain). All organizations should use the Suichies Model, or a similar model, to analyze whether or not to use the blockchain.[225]

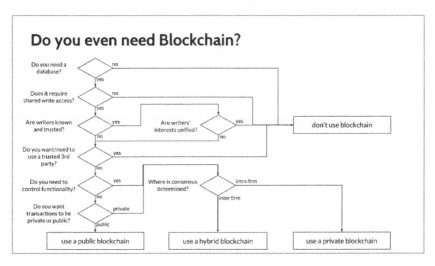

Figure 21 Suichies model; courtesy of Brad Suichies

DEPARTMENT OF HOMELAND SECURITY MODEL

The Department of Homeland Security (DHS) Model is extremely important because DHS is one of the largest U.S. government agencies and is primarily responsible for national security. DHS found blockchain technology was important enough to create a flowchart. The DHS analysis of data, data stores, data records, and sensitive identifiers is also important because the flowchart recognizes the best option may be to use different kind of databases (standard, managed, or encrypted) rather than a blockchain. Finally, the DHS flowchart considers the presence of sensitive information and comes to the logical conclusion that an encrypted database is more functional for sensitive information than a public blockchain.[226]

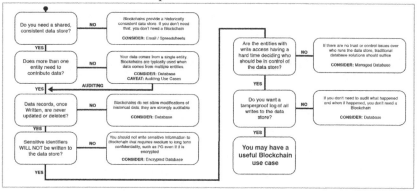

Figure 22 DHS blockchain flowchart analysis

Blockchain Technology and Government Contracting

The DHS flowchart shows that federal agencies are paying attention to blockchain technology, but DHS is only one agency. Additional agencies recognize the critical role blockchain technology can play in the business of managing the U.S. government. The General Services Agency (GSA) website, GSA.gov, lists blockchain technology uses for government agencies including:

- Financial management
- Procurement
- IT asset and supply chain management
- Smart contracts
- Patents, Trademarks, Copyrights, Royalties
- Government-issued credentials like visas, passports, social security numbers, and birth certificates
- Federal personnel workforce data
- Appropriated funds
- Federal assistance and foreign aid delivery[227]

This is an impressive list and combined, all of the listed areas impact every government agency. GSA recognized the importance of blockchain technology and created GSA's Emerging Citizen Technology Office launched the U.S. Federal Blockchain Program. The Federal Blockchain Program coordinates programs with federal agencies and U.S. businesses interested in exploring distributed ledger technology and its implementation within the government.

GSA also hosted the first U.S. Federal Blockchain Forum in July 2017. The Federal Blockchain Forum united over 100 federal managers to discuss blockchain use cases, limitations, and solutions. This information is important for businesses as the federal government is a major purchaser of products and services. According to the March 2017 Government Accountability Office (GAO) Contracting Data Analysis, in fiscal year 2015, federal agencies obligated over $430 billion in contracts for products and services.[228] Federal government contracts are a major opportunity for businesses and will soon be a major opportunity for blockchain businesses.

The private sector is paying attention to government blockchain spending as well. In "Opportunities Show Growing Agency Use of Blockchain," *ITCON.com* reports "federal agencies have funded $8.4 million in blockchain contracts since fiscal 2015." But more than half of these contracts were awarded after July 2017.[229] Cointelegraph.com quotes the 2019 IDC Government Insights Report (hereinafter "The IDC Report") forecasts that U.S. government blockchain spending is expected to increase to $123.5 million by 2022. The IDC Report also projects blockchain spending is forecasted to grow from $4.4 million in 2017 to $48.2 million in 2022.[230] Those figures represent a 1,000% increase in spending.

Another U.S. government agency, Department of the Treasury, Bureau of the Fiscal Service created an analysis to determine if a business needs a blockchain. A direct quote of their analysis follows:

1. **Determine if Blockchain is a Good Fit.** Developing a deep understanding of the problem you're trying to solve is an important first step for deciding whether blockchain is the right tool.

2. **Understand Pain Points (and Good Points).** Start by identifying and interviewing a broad group of stakeholders who, through the process, will help uncover "pain points" that may have otherwise gone undetected.

3. **Map Business Processes.** Considerable time goes into mapping your current processes and identifying friction points—places within a process that are costly, time consuming, manual, or otherwise inefficient.

4. **Build a Diverse Project Team.** Putting together the right project team is important. Having cross-functional representatives, and especially your IT shop (blockchain is, after all, software), helps to develop a better, more robust blockchain application.

5. **Consider Agency Governance.** Agency governance processes can be time consuming. Preparing for and presenting your project in front of your investment review and technical review boards can bring to light perspectives or issues considered.[231]

The Department of Treasury analysis is a good starting point if your company is considering using blockchain technology in their business implementation. As we are discussing blockchains and governments, the last component, "agency governance" merits further discussion.

BLOCKCHAIN TECHNOLOGY AND COMPLIANCE REQUIREMENTS

Businesses with compliance requirements have additional considerations when using blockchain technology. First and foremost, these businesses must comply with any and all laws and regulations related to their business, which means use of blockchain technology cannot jeopardize compliance. In this scenario, businesses have three options:

1. Continue to research blockchain technology and wait for the government agency to use blockchain technology in their compliance efforts.
2. Incorporate blockchain technology in non-compliance areas of their business.
3. Create a redundant blockchain system where the current system remains in operation to ensure compliance and the blockchain system is used with the same data to determine the functionality in a test case environment.

These three scenarios are solutions that create systems where blockchain technology can be used while compliance is maintained.

HOW TO INCORPORATE BLOCKCHAIN INTO YOUR ORGANIZATION

Although hype surrounds both the potential and benefits of blockchain technology, it may not be the best use of your company's time and resources, especially this early in the use of the technology. Thus far this chapter addressed the question, "should I incorporate a blockchain into my business?" If one or more of the resources answered the question for you in the affirmative, then you will benefit from the next section, which addresses the question, "how should I incorporate a blockchain into my business?" In addition to the Department of Treasury of Financial Services quoted above, consider the following recommendations when considering the use of blockchains in your business.

1. DETERMINE THE CHALLENGE A BLOCKCHAIN WILL SOLVE.

This takes us back to the quote at the start of the chapter about using a blockchain to solve problems, not to find problems. Does your industry or do your clients have serious challenges or business practices that current technology cannot solve? As discussed in Chapter 6, health care, cloud storage, and human resource challenges are examples of how blockchain technology can resolve them. Every business and industry has challenges in need of solutions. The questions are, what are the challenges and how does a blockchain solve them?

2. ESTABLISH A "BLOCKCHAIN TEAM/WORKING GROUP."

Create a working group to research the best ways to solve the challenges. The Blockchain Team should review the organization business plan and other strategic plans. Blockchain Team/Working Groups should look for legitimate and serious problems facing the organization and determine how incorporating blockchain technology can solve organizational problems. As discussed in Chapter 6, legislatures use the same approach by creating working groups and legislative studies to determine the impact of a blockchain on their governance models and constituents.

The Blockchain Team should have an experienced programmer or architect employee. A full understanding of functionality and capacity of a blockchain is critical to successful integration. If your company does not have an experienced blockchain programmer, consider using a blockchain consultant to fill the experience gap.

3. CHOOSE THE FIRST BLOCKCHAIN PROJECT.

The first blockchain project should be a small project netting very little risk. Consider using a test project based on a past or present business challenge. Unlike using a blockchain to solve a current business challenge, in the test project, the company is in full control of all of the variables. Therefore, if a blockchain application fails, the business or client will not lose any money or business.

4. CREATE METRICS.

Metrics are an important way to determine if any newly implemented procedure or strategy is successful, including blockchain integration. Create reasonable and achievable short-term metrics to measure the success of the incorporation of blockchain technology. As mentioned in my first book, *Diversify or Die*, metrics should be established according to the SMART acronym.

S—Specific (clearly specify the intended outcome/s). Examples of potential outcomes from blockchain technology including: measurable cost reduction, successful fundraising efforts, improved data and security, updated payment systems, etc.

M—Measurable (quantifiable; stating exactly what the criteria for success are and how they will be assessed/measured). Building off of the above category, measurable goals can be added to determine: the target amount of cost reduction, a fundraising goal, a target to determine successful data and security, or the amount of users by number or percentage for a new or updated payment system. These measurements will be an important step in determining the success of the project.

A—Attainable (challenging but achievable; employee has the skills, time, resources, and authority to deliver the expected results). Attainability is an important metric. As blockchain technology is still very new, it is important not to try to attempt to achieve too much too soon. Balance your resources required to maintain operations with the resources required to complete a blockchain project.

R—Realistic (relevant to the employee's role; is willing and able to work toward its achievement). After considering attainability, employers need to consider the realistic impact of using resources to complete the blockchain project. Keep in mind that what seems attainable in theory, may not be attainable when all factors related to running a successful business are in play.

T—Time-bound (clearly defined time frame or target date). The two most important factors in the time bound metric include external deadlines for clients or investors and meeting self-imposed deadlines. If your business has the luxury of time to complete a blockchain project, consider implementing a long-term deadline.[232]

5. INTERFACE WITH INDUSTRY SPECIFIC BLOCKCHAIN ORGANIZATIONS.

Another step is to join a blockchain organization. Blockchain organizations create opportunities to interface with other industry professionals and to learn from each other. If your industry does not have a blockchain organization, consider starting one and becoming a blockchain pioneer in your industry.

ACTUAL EXAMPLE: PUBLISHING *BLOCKCHAIN OR DIE* ON THE BLOCKCHAIN

I decided to incorporate blockchain technology into my publishing business and book sales. Since book publishing is not a pioneering blockchain industry, initially I did not see an opportunity to incorporate blockchain technology into my business. After I used the steps outlined in this chapter, I realized I had a rare opportunity to publish *Blockchain or Die* on a blockchain. The thought of publishing *Blockchain or Die* on a blockchain opened the floodgates of possibilities. As very few books have been published on a blockchain, publishing a book about cryptocurrencies and blockchain technology on an actual blockchain would be a perfect fit.

The following steps provided the solution to incorporate blockchain technology into Better ME Better WE Publishing:

1. DETERMINE THE CHALLENGE A BLOCKCHAIN WILL SOLVE.

The publishing industry has changed at a rapid pace since the commercial use of the internet. Publish-on-demand and peer-to-peer file sharing dramatically decentralized and changed the publishing industry forever. As stated in Chapter 1, *Blockchain or Die* addresses the problem that people do not know about cryptocurrencies or blockchains. The basic publishing business model is to sell books with good content to as many people as possible, and the number of ways to publish a book has increased with new technologies.

Once published, successful authors use their book as a platform for book signings and public speaking engagements. If the book is business, science, or policy related, the author can use his/her book to build a consulting business.

2. Establish a "Blockchain Team/Working Group."

Better ME Better WE Publishing did not need an internal "blockchain team." Instead, we created an external blockchain team by connecting with blockchain pioneers in the publishing industry. There are very few blockchain publishing organizations, but one blockchain publishing company met all of my blockchain publishing needs: Publica. Publica is bringing publishing to blockchains by hosting books, providing a platform for presales crowdfunding, and creating a blockchain e-commerce solution. I decided to use Publica to publish *Blockchain or Die* on a blockchain.

3. Choose the First Blockchain Project.

Publishing *Blockchain or Die* on a blockchain was the first blockchain project, but I also published *Blockchain or Die* through traditional publishing methods. A critical component was to determine the method to accept payment for blockchain sales. This led to asking the following questions: What is the best blockchain payment system? Which cryptocurrencies should Better ME Better WE Publishing use for payment? Is there a method to have *Blockchain or Die* purchasers use credit cards, debit cards, PayPal, or another means of payment on a blockchain?

As discussed earlier in this chapter, any payment system must be user friendly. Amazon.com is extremely user friendly and facilitates the purchasing and delivery of books. Blockchain publishers and booksellers must provide a similar user-friendly purchasing experience to compete with Amazon and other online book retailers.

4. CREATE METRICS.

Book sales are the most important metric related to the success of a published book, including books sold on a blockchain. Other metrics include the number of book signings, paid public speaking engagements, and consulting jobs obtained from *Blockchain or Die*. The application of the SMART follows:

Specific—One of the goals is to create a quality book that will achieve best-seller status. Since *Blockchain or Die* was published on both traditional publishing platforms and a blockchain, there will be two sets of sales metrics to determine success: traditional publishing and blockchain publishing.

Measurable—Book sales are clearly measurable on both traditional and blockchain publishing platforms.

Attainable—Robust book sales to obtain best-seller status is attainable.

Realistic—Selling enough books to achieve best-seller status with a traditional publisher is a realistic objective; however, metrics to determine best-seller status on a blockchain has yet to be determined.

Time-bound—Most book sales take place within the first year of publication. Strategically, the most important time to plan and execute the *Blockchain or Die* blockchain publishing strategic plan is before *Blockchain or Die* is released. The number of a book's pre-sales is important to the successful release of a new book. More importantly, since so few books have been published on a blockchain, promoting *Blockchain or Die* as well as promoting blockchain publishing, and the new business model for publishing, adds additional marketing value to *Blockchain or Die,* Better ME Better WE Publishing, and branding myself as a blockchain innovator.

5. INTERFACE WITH BLOCKCHAIN ORGANIZATIONS IN YOUR INDUSTRY.

Many blockchain organizations serve as an excellent resource for information, contacts, and training. The organizations I chose to join include:

- The Government Blockchain Association (GBA)
- Blacks in Blockchain
- Diversity in Blockchain
- Many LinkedIn Cryptocurrency and Blockchain Groups

Prior to joining these organizations, I compared the membership dues to the value provided by the organization. These organizations provided excellent resources and contacts for the dues required. I constantly review the membership value of other blockchain organizations as part of my ongoing blockchain journey.

BLOCKCHAIN CERTIFICATION PROGRAMS

Once a blockchain is adopted as a business practice, leadership and employees should receive blockchain training. Blockchain certification programs are an excellent way to learn about blockchain technology. Many organizations provide blockchain certification courses, but not all blockchain certification programs are created equally.

As with any other training program, some certification programs are solely designed to profit the training company, not to provide an outstanding training experience. As a Certified Training Director with decades of experience of designing and delivering many certification trainings, I created a checklist to determine the credibility and quality content of a blockchain certification program.

1. Experienced Instructors: The first criterion is to select a program with experienced instructors to conduct the training in a competent manner.

2. Length of the Certification Program: Blockchain technology is a new and complicated topic, so there is a lot of material to cover. One-day certification programs will not provide sufficient learning time and depth to understand the complexities of blockchain technology. Multi-day certification programs provide more time to cover the essential blockchain topics. The only exception to the one-day training is recertification training.

3. Choose a Highly Interactive Certification Training Program: Whether the training is classroom or online, interactive trainings provide the most desirable training environment for the adult learner. The lecture model for training is outdated and, frankly, quite boring. The dynamic nature of blockchain technology in an interactive training provides an excellent opportunity to deliver exciting and groundbreaking content. In an interactive training, participation is enhanced by well-designed and implemented training scenarios, which provide a meaningful and lasting training experience.

4. Certification Test: A good certification training program has a certification test at the end. Certification tests may not seem desirable, but they serve as quality control for certifications. Once the blockchain community sees the quality of the certificated trainers, the training program will be highly valued and the certified trainers will have sought after status.

5. Professional Support: If you desire a blockchain related career, it's best to choose a certification training program with a robust professional support system. Professional blockchain support systems include the following offerings:

a) Regular meetings

b) Affiliation with other organizations

c) Hosted blockchain conferences, meetings, or hackathons

d) Maintained job board

This information will be very helpful in choosing a certification training program to start or advance your blockchain career. Use this list to choose the certification program that best suits your needs.

SUMMARY

The best way to end this chapter is by revisiting an important section of the quote that started this chapter: "We need [blockchain systems] to be user friendly." This chapter focused on determining if your business needs a blockchain, and if necessary, successfully incorporating blockchain into your business. Every blockchain model that does not use sensitive data should include a blockchain solution to create user-friendly products and services that solve critical business challenges.

SUCCESS STRATEGY ACTION ITEMS

1. Continue to research the status of blockchain technology in your industry or any industry of interest. This information will be rapidly evolving so frequent status checks are recommended.

2. Choose a low risk blockchain project for the first project. After reading this chapter, you may have considered one or more blockchain projects that could solve a challenge for your business. Choose a blockchain project that will not disrupt your business, especially if the project is not successful.

3. Apply the SMART model to incorporate blockchain technology into your business.

4.   Research the legislation for blockchain technology in the states and locales you conduct business and determine how it will impact your business.

5. Join a blockchain organization. Reading books, articles, and blogs act as excellent resources, but the best way to complete your blockchain experience is to engage with other blockchain enthusiasts. The more involved you are, the more you will learn.

6. Research legal counsel with expertise in blockchain technology. Legal organizations, such as the American Bar Association or state bar associations are a good place to start. Also, some law firms now specialize in blockchain technology and advertise their blockchain legal services on the internet.

7. If you are interested in publishing a book, consider publishing on a blockchain. The information in this chapter provides an excellent start. As of the time of publication, blockchain is still a new business concept so publishing on a blockchain at this point in time will make you an "innovative author."

8. Research blockchain certification training programs. If you want to become a certified blockchain consultant or trainer, consider completing one or more blockchain certification training programs.

[9]

TEN WAYS TO MAKE MONEY WITH BLOCKCHAIN TECHNOLOGY TODAY

"When anyone can use blockchain technology to make money as they would currently use Amazon, Airbnb, eBay, Uber, and Upwork that is when the average person will be able to use a blockchain to make a living."

—Eric Guthrie, Esq.

THIS QUOTE REPRESENTS the next step in user interface. Amazon, Airbnb, eBay, Uber, and Upwork are five very different companies with three things in common. They allow (1) average people to (2) use the internet to (3) make money. I wrote this chapter for the reader who is not interested in investing in cryptocurrencies, using a blockchain in their business, or creating a blockchain company.

Starting a blockchain-based business or incorporating blockchain technology into your existing business are two ways to profit from the blockchain, but there are other ways to profit. This chapter provides ten ways to make money using blockchain technology. Some of the methods in this chapter were generally discussed in previous chapters, and others are presented for the first time. All of the actions addressed in this chapter have one thing in common; they do not require a great deal of technical expertise.

THE BLOCKCHAIN ECONOMY

Globally, tens of millions of people have extra jobs or "side gigs" in this gig economy. The statistics for Amazon, Airbnb, eBay, Uber, and Upwork show the impact of the gig economy.

- In 2017, *Marketplace Pulse* reports Amazon has five million sellers across all of the Amazon Marketplaces (USA, UK, Germany, France, Canada, Japan, India, Italy, Mexico, Brazil, and China).[233]
- In 2018, Airbnb reports six million listings in 191 countries and 181,000 cities.[234]
- In 2017, *Small Business Trends* reports eBay has 6.7 million sellers and has sold products in 190 global markets.[235]
- In 2018, Uber reports two million drivers globally and 750,000 in the United States.[236]
- In 2017, Wikipedia reports Upwork, formerly Elance/ODesk, has twelve million registered freelancers and five million registered clients.[237]

These reports show the impressive numbers of people who are taking advantage of the gig economy. The recent investment and industry trends set forth in this book show how the blockchain economy is the next global phenomenon.

The actions in this chapter probably won't make you millions, but they provide exposure to blockchain technology while providing profit for your time and efforts. Some of the actions can be accomplished in a day, others may take weeks or months, but all can be achieved without blockchain programming.

Opportunities to profit from blockchains present themselves in many ways, including individual freelancing opportunities, renting your property, driving for a rideshare company, and using DApps to invest in cryptocurrencies. But before we review the ten ways to make money in blockchain today, let's look at buyer beware from this perspective.

Buyer beware was introduced with ICOs in Chapter 4. Many of these ICOs raised millions of dollars, but some were scams requiring SEC enforcement. Scams come in many shapes, sizes, and dollar amounts, including small-scale blockchain opportunities.

The ten ways to make money with blockchain technology today follow.

ONE: INVEST SMALL CHANGE

Technology simplified the investment process and the use of apps for investing is the latest innovation partnering investment and technology. Traditional apps, for example Acorn and Robinhood, make investing more affordable by reducing the amount of money required to invest in the stock market from dollars to cents.

In similar fashion, blockchain investing apps are making it easier to invest in cryptocurrencies. All of the cryptocurrency exchanges mentioned in Chapter 1 have apps, but do not provide the option to invest small change. Bundil and Coinseed are two apps that allow users to invest in cryptocurrencies with spare change.

Bundil, as seen on Shark Tank (season 10 episode 3), rounds up your purchases to the next dollar and invests the spare change in cryptocurrencies.[238] According to the Bundil app, Bundil users who spend \$3.62 will have \$0.38 invested in cryptocurrencies totaling \$4.00. Cryptocurrency investors can also use Bundil to purchase as many coins as they desire. The Bundil app allows users to track their portfolio and monitor their progress.[239]

In similar fashion to Bundil, the Coinseed app allows users to convert their spare change into cryptocurrencies. Coinseed provides a number of products to simplify the blockchain investment experience.

- Crypto Cash Back—Earn crypto cash back when you shop at your favorite shops and restaurants.
- Crypto Portfolio—Create your own unique portfolio from dozens of cryptocurrencies.
- Round-Ups—Invest your spare change from everyday purchases.
- Recurring Investments—Set up daily automatic investments of \$.05—\$5 and relax.
- One-Time Investments—Just invest as little as \$5 and enjoy seeing how it grows.
- Auto Rebalance—Enable the auto rebalancing feature and benefit from price fluctuations.
- Play Mode—Don't want to risk real money? Start with play mode and hone your skills.
- Weekly Tournament—Enter our weekly simulation tournaments and win big money.[240]

Both apps are very user friendly and present cryptocurrencies in non-technical language with easy steps to set up an account.

TWO: STAKING

Staking is another way of earning cryptocurrencies. Chapter 3 covered "buy and hold" as a cryptocurrency investment strategy and staking is a method to earn a dividend while you hold onto your cryptocurrency. In staking, cryptocurrency purchasers buy and hold crypto-coins in a special wallet to earn regular dividends or profits from the cryptocurrency.

Proof of stake cryptocurrencies, covered in Chapter 5, usually use staking, aptly called a staking reward, as a way to reward purchasers of their coin and secure the blockchain as well. As of the time of this publication, the following coins offer staking as a dividend:

- Dash (DASH)
- NEO (NEO)
- PIVX (PIVX)
- Reddcoin (RDD)
- Neblio (NEBL)
- Qtum (QTUM)
- OKCash (OK)
- NAV Coin (NAV)
- Stratis (STRAT)

Finally, it is important to note that purchasers must keep their wallets online most of the time to earn dividends.[241]

THREE: EARN BITCOIN BY SELLING STUFF ON BLOCKCHAIN SITES

Selling stuff on a blockchain site is pretty straightforward. In a similar fashion to eBay and other online auction sites and retailers, anyone can buy goods or services on a blockchain. Blockchain eBay equivalents include Bitify and Purse.io. Bitify allows users to buy and sell goods on their blockchain.[242] Purse.io helps Amazon sellers sell their goods for a discount on its blockchain platform.[243]

Lack of exposure is the biggest issue with selling goods or services on a blockchain site, as most people do not know about the blockchain sites that sell goods and services. When the average person starts to use blockchains on a daily basis, buyers and sellers will transact more blockchain-based sales.

FOUR: MINING AS A SERVICE

Mining was the first way to make money in the bitcoin blockchain. When miners realized the amount of profit in bitcoin mining, they created specialized computer systems, Application-Specific Integrated Circuits (ASIC) circuits, and mining rigs with hundreds of computers to mine bitcoin. Mining operations requires technical expertise, a great deal of investment for the necessary equipment, and the power to run the equipment. Mining has shifted from conventional GPU set-ups to ASICs to Field Programmable Gate Arrays (FPGAs).

The best days of bitcoin mining may be in the past as bitcoin is steadily decreasing the payout to miners. Decreased payout translates into decreased profit. Companies have devised a business model, mining as a service, (MaaS) as an alternative to traditional mining.

MaaS is an arrangement where you use a mining company's computers and software to potentially produce cryptocurrency for your use. MaaS is the same as any other software as a service arrangement where you are simply renting/leasing the means to accomplish an end result.

CoinDesk's "How does Cloud Mining Work" provides an insightful view of cloud mining pros and cons.

> **Pros:** You don't have to buy computers that will run twenty-four hours a day, add heat to your house, run noisy fans, and increase your electric bills.

Cons: As a "cloud miner," you don't have control or any inside information about the company's operations, so potential fraud is an issue. Also, you have to share in the profits, which you receive after the cloud mining company takes their share. Your profits change as the cryptocurrency you are mining fluctuates in value.

Three MaaS companies include: Nuvoo Mining, DMG Blockchain Solutions, and Mining Buster. Each MaaS company website has a lot of information and readers are encouraged to review each website and research the MaaS company before choosing a cloud mining company.

FIVE: BLOCKCHAIN CLOUD STORAGE

Blockchain Cloud Storage was covered in Chapter 6, but it is important enough to mention again. As a reminder, blockchain cloud storage computer owners "rent out" extra hard drive space on their computer. This presents an opportunity for computer owners to profit from blockchains. Chapter 6 covered Sia and Storj, two blockchain storage companies, but State of DApps showcases more blockchain cloud storage companies, including XCloud and EOS File Storage. These companies are using blockchain to evolve cloud storage.

SIX: BLOCKCHAIN SOCIAL MEDIA SITES

Similar to YouTube and Facebook revolutionizing the way we view content and videos, LBRY, Steem, and DTube has revolutionized content by providing blockchain based social media. On blockchain social media sites, content creators are compensated in tokens for posting content on blockchains. According to LBRY.io:

"LBRY is a free, open, and community-driven digital market-place that enables content sharing, monetization, discovery, and consumption. Publishing in LBRY is the process of sharing your content on the network. You have the ability to set the price per view (can be free too)[,] which is paid directly to you. LBRY users can also play arcade style videogames and developers can even help build out the LBRY system. LBRY users even receive awards for bring new users to the LBRY blockchain."[244]

Will these blockchain social media sites replace existing social media sites? It's too early to tell, but the blockchain business model to award content creators with rewards and bringing new users to the site are two reasons to use these new sites.

SEVEN: AIRDROPS

Historically, airdrops date back to World War II when the military dropped supplies from airplanes to troops when they could not land planes. In the blockchain world, "airdrops" are where blockchain companies "drop" a few free crypto tokens to parties that performed marketing tasks for the tokens, including: liking, following, and posting on social media. Once the marketing tasks are completed, the crypto tokens are deposited into the airdrop users' wallet, completing the airdrop transaction.

While a few free crypto tokens may not seem like a lot of money, if they increase in value, the small airdrop will increase in value as well. As covered in Chapter 3, the original price for bitcoin was .003 and it increased to a high of $19,666.[245] Although bitcoin was not airdropped, imagine if you signed up for an airdrop for a crypto token that increased 1/10 of the value of bitcoin when bitcoin was valued at .003 cents? Still a very impressive increase.

Quality over quantity is critical when choosing airdrops. Unfortunately, many airdrops are scams designed to collect your email address and other personal information. For the security of your information, carefully review the airdrop information before you sign up for the airdrop. The following questions will help in your airdrop decision process:

1. What is the business model for the crypto token and does it make sense? Don't rely on the airdrop marketing hype. Instead, carefully consider their business model. One approach is to use the information in Chapter 8.

2. Does the airdrop have a professional and credible website? A company's website is one of the first resources investors research when deciding whether or not to invest in a blockchain project. Serious blockchain companies invest in a very professional website to ensure investors receive a good first impression and enough information and research to make a sound investment decision.

3. Does the airdrop announcement use proper spelling, grammar, and pronunciation? Similar to spam mail, poor spelling, grammar, and pronunciation are common signs you are reading a "spam scam," and the same principles apply to airdrop announcements.

4. Are the executives and developers behind the coin or token credible? Do they have blockchain or cryptocurrency experience or experience in a related field? These questions are critical to deciding whether or not to sign up for the airdrop. Anyone can write a white paper, create an airdrop communication, and market their blockchain product as an airdrop in the hopes of building the hype to sell their cryptocurrencies or tokens on the market and make a profit. One way to avoid "airdrop scams" is to research the credibility and experience

of the executives and developers. Accurate airdrop information will give you a better idea of the credibility of the entire blockchain project.

5.   What are blogs reporting on the airdrop? Blogs are an excellent resource to research the quality of an airdrop. Experienced bloggers conduct detailed research and are up to date on current events. If the airdrop is a scam, related blogs will eventually post their negative airdrop experience. Once the airdrop scam hits enough blog posts, the number of airdrop subscribers will likely decrease.

The airdrops that pass these questions are worth considering. Legitimate coin creators are looking to spread the word about a credible crypto token to create demand. For example, the Investinblockchain.com article, "Three Crypto Airdrops In Q3 2018 You Should Know About" lists FXPay (FXP), BigBang Token (BBT), and HireVibes (VBT) as three airdrops offering the best rewards compared to the required marketing tasks.[246] Two examples of airdrop resources include Airdropalert.com and Topicolist.com/Airdrops.

EIGHT: BLOCKCHAIN BLOGGING

Similar to blogging for money on the internet, bloggers can blog for money on blockchains. There is a great deal of information online on creating a successful blog, but the some of the major steps include:

1.   Set up your blogging platform.
2.   Choose one or more income streams.
3.   Create quality content.
4.   Market your blog.
5.   Engage your readers and membership.

NINE: BLOCKCHAIN FREELANCING

The internet provides technology for freelancing on a global level. Numerous websites, including fiverr.com and freelancer.com give sellers the platform to post specific assignments for freelancers to complete for compensation. "Freelancing in America," published by Edelman Intelligence and Commissioned by Upwork and Freelancers Union, made the following key findings:

- There were 57.3 million people who freelanced in 2017.
- The freelance workforce grew at a rate of three times faster than the U.S. workforce overall since 2014.
- At its current rate of growth, the majority of the U.S. workforce will be freelancers by 2027.[247]

Freelancing has become a major component in the "gig economy," but the next economy may be the "blockchain economy."

Freelancing already exists on blockchains. The freelancing model is the same as the internet model, but there are a few major differences.

First, as with all blockchain technology, blockchain freelancing is peer-to-peer. Currently, freelancing companies charge the purchaser a percentage of the money off the top for freelancer's work. When a purchaser agrees to pay the freelancer $100, the purchaser pays the freelancing website company as much as 20% to use their platform. In this scenario, the total cost to the purchasers is $120. This is a sizable portion of the money to the centralized company instead of the freelancer.

Second, blockchain freelancers are required to submit verifiable proof of their experience and expertise. Current freelancing sites do not require freelancers to submit verifiable proof of expertise.

Third, blockchain freelancers are paid in cryptocurrencies, usually bitcoin or ether.

Finally, because the assignment is created and administered on the blockchain, the freelancer is paid immediately after they complete the assignment. This is the same concept as the digital vending machine discussed in Chapter 5.

As a blockchain instructor, attendees often comment they cannot freelance because they are not blockchain programmers. A common conversation with freelancers goes as follows:

"I am interested in blockchain freelancing, but I am not a programmer," a freelancer shared with me.

"There are many other blockchain freelancing opportunities besides programming."

"Really? I have only heard about blockchain programming."

"Blockchain companies need marketing experts, graphic designers, lawyers, and writers just to name a few. See for yourself." I used my smart phone show them postings on a blockchain freelancing site.

"Wow," said the freelancer. "I had no idea. How do I start?"

"Set up an account on a blockchain freelancer site and create a profile that highlights your qualifications and expertise."

"Great, thanks!"

After attendees see the actual postings for blockchain assignments, they realize sellers are looking for many other areas of expertise beyond blockchain programmers. Two blockchain freelancing sites include: https://www.reddit.com/r/Jobs4Bitcoins/ and https://ethlance.com/.

If you don't have a lot of time to work on blockchain projects, microtasks are another option. Microtasks are a form of freelancing, but there are a few differences. As the name indicates, "microtasks" are usually shorter in time and complexity and pay less than freelancing. Coinbucks.io states Coin Bucks is a "bitcoin CPA affiliate network that allows you to earn bitcoin by completing online promotional offers."[248] Once you complete the promotional offer, the bitcoin is immediately deposited into your wallet. The more you use Coin Bucks, the more bitcoin you will earn.

TEN: BECOMING A BLOCKCHAIN TRAINING PROVIDER

Chapter 8 covered training certification courses; however, this section will cover profiting by providing training as a blockchain training provider. Globally, blockchain training providers are needed to educate businesses and individuals on the uses, benefits, and risks of blockchain technology. Although there are many blockchain training providers, there is always a need for experienced ones.

Becoming a qualified blockchain training provider is covered in the last section of this chapter, as it takes time and effort to become a blockchain training provider. If you are new to blockchain technology, the best way to become a qualified training provider is to complete a blockchain trainer certification course. Once certified, and blockchain trainers have access to the training material, they can market and profit from their training expertise.

Blockchain trainers can use the following activities to grow their training business:

1. Research and read everything possible about blockchain technology. It is important to understand the trends, issues, challenges, solutions, and new applications and technology.

2. Join LinkedIn Learning and Development Groups and other professional development groups.

3. Publish Articles. Publishing articles and becoming a thought leader is one of the best ways to attract new customers.

4. Get involved with blockchain organizations. Serve on steering committees, coordinate events, and volunteer to provide blockchain courses. Sign up with organizations like Government Blockchain Association. The more responsibilities and assignments you accept, the more you build your experience. If you are not confident enough to provide training, co-train with an experienced trainer until you are comfortable enough to deliver the training. Shadowing a trainer is also an opportunity to develop a mentor–mentee relationship.

5. Complete Certification Courses. Classroom or online-based certification training increases your blockchain knowledge and increases your credibility.

6. Monetize Your Trainings. Designing training programs will not provide any financial benefit if clients do not buy them. Conduct market research and determine the best price point for your training program.

7. Market Your Trainings. Determine your client base and the best way to expose them to your trainings.

Even if you do not want to become a blockchain trainer, completing a blockchain certification course may boost your career as employers are looking to hire blockchain experts. Your certification may be the cornerstone to a new career with your current or future employer.

SUCCESS STRATEGY ACTION ITEMS

In a way, this chapter is a long list of action items, but here are some recommendations on how to proceed from here.

1. Choose one or two of the categories based on your experience or interest level for your areas of focus. Don't try to focus on too many areas at the same time.

2. Choose categories in your budget as some require investment. Mining and becoming a certified trainer, for example, require a financial investment while freelancing and blogging can be initiated with little to no money.

3. Find someone who is successful in the areas of interest and learn from their experience. You don't have to do this alone. There are many people in the blockchain community who are more than willing to share their knowledge and experience. You can find them online or in one of many blockchain organizations.

4. Give yourself time to succeed. If you don't start making money in a few days or a week, don't give up. The Egyptian pyramids were not built in one day, but the first brick was laid on the first day. This is what this chapter is about, laying your first blockchain brick, and building your blockchain empire from here.

5. Do not limit yourself to the ten categories in this chapter. There are many other ways to make money with blockchain technology, enough to fill up a book. If you know of any other blockchain entrepreneurial ideas, pursue that idea. Conduct your own research, talk to other blockchain experts, or better yet, be an innovator and create one of your own.

[10]

CRYPTOCURRENCIES, BLOCKCHAIN TECHNOLOGY, AND POSITIVE REAL-LIFE IMPACT

"Western Union spends and earns billions to do what bitcoin does for free. Instead of Western Union, migrant workers (or businesses operating on their behalf) could use bitcoin to send payments from one country to another through email, without worry of fraud or needing to support an elaborate exchange or credit market. It would be real-time, immediate settlement at a fraction of the cost. In ten years, instead of international drugs, bitcoin could act as a genuine lingua franca for international work."[249]

—Timothy Carmondy, National Geographic

MOST CONTENT, including the content in this book, about cryptocurrencies and blockchain technology focus on currency, investment, and business applications. However, one of the important aspects of blockchain technology is the powerful impact on communities in greatest need of access to financial institutions. As stated by Timothy Carmondy, migrant workers' use of cryptocurrencies provides more secure and affordable options for payment after a hard day's work, and those options give them greater control of their basic necessities.

POWERFUL TECHNOLOGY, POWERFUL IMPACT, POWERFUL STORIES

As mentioned in the introductory chapter, there is a massive amount of information online with the most amazing stories covering how this new technology is impacting people around the world. All of the stories were in individual articles, so interested parties would have to research and read many articles to understand the global scope, depth, and impact of this technology, especially in developing countries.

The information in this chapter shows the real value of cryptocurrencies and blockchain technology. There are many examples of technologies that make the rich richer and the poor poorer. Blockchain technology has societal, financial, and technological components that operate together to empower people.

Societal—Most blockchain technology is open source, which puts the technology in the hands of the people rather than governments and corporations that seek to control or profit from the technology.

Financial—Blockchain technology provides unbanked populations the means to open an account and to manage their financial affairs without the government or banking requirements that many cannot meet. This gives the global unbanked population the means to save money, invest, and transact business for the first time.

Technological—As stated in the Societal section, most blockchains are free and open source so it is available to assist anyone who can use the technology so they level the technology gap.

All of these components were discussed in earlier chapters, but in this chapter they will be discussed in the context of improving the lives of the underprivileged.

On a global scale, the true power and benefit of cryptocurrencies and blockchain technology is to move millions out of poverty and to create a wealthier global society. In this new wealthier society, financial stresses on governments to cover the costs of the underprivileged will decrease as they will have access to digital identities, property records, affordable financial transactions, and a new found earning and spending power that will add to local economies and the tax base.

However, empowering the underprivileged is not good news for everyone. Governments with a history of unfair policies and financial practices to oppress the underprivileged will not embrace this new technology. In fact, they may see it as a threat to their oppressive status quo. Also, intermediaries with a history of over-charging the underprivileged for basic services, including banking and currency transactions, will also see this technology as a threat to their status quo. For example, remittance businesses are a major example of intermediaries charging the underprivileged for basic monetary transactions.

IT COSTS $15 BILLION TO SEND MONEY TO FAMILY AND FRIENDS

For most of the people reading this book, access to money is never an issue. It's easy to go to a bank, show two forms of government identification, give a local address, and open a savings or checking account. That is the life of the "financially privileged."

As the son of immigrants with strong ties to the immigrant community, I know many of the issues immigrants face, especially when it comes to financial issues. In many circumstances, one or more members of an immigrant family enters the United States to find work, mostly because they cannot find living-wage work in their home country. In this situation, the migrant employee is primarily working in order to send money back to their families in their home country.

According to the January 2018 *Pew Research Center Report*, $138,165,000,000 was sent from the United States to other countries in 2016. Migrant employees sent an estimated $574 billion to relatives in their home countries.[250] Here are the top twenty countries that received remittances in 2016 from migrant employees.[251]

1.	Mexico	$30,019,000,000
2.	China	$16,141,000,000
3.	India	$11,715,000,000
4.	Philippines	$11,099,000,000
5.	Guatemala	$7,735,000,000
6.	Vietnam	$7,725,000,000
7.	Nigeria	$6,191,000,000
8.	El Salvador	$4,611,000,000
9.	Dominican Republic	$4,594,000,000
10.	Honduras	$3,769,000,000
11.	South Korea	$2,834,000,000
12.	Germany	$2,801,000,000
13.	France	$2,373,000,000
14.	Thailand	$1,859,000,000
15.	Jamaica	$1,800,000,000
16.	Colombia	$1,767,000,000
17.	Japan	$1,593,000,000

18. Haiti	$1,494,000,000
19. Pakistan	$1,323,000,000
20. Italy	$1,314,000,000
Total	$122,757,000,000

Combined with the information from the World Bank Remittance Worldwide Report, which reports global remittances cost an average of 7.13 percent, of the $122,757,000,000 of the top twenty countries, an estimated $8,752,574,100 profited remittance organizations rather than the remittance recipients.

Finally, the remittances in this discussion take place under regular circumstances; but remittances become a critical lifeline in a disaster or crisis situation. The companies ConsenSys, Maker, and Dether teamed up to build Project Bifrost, a system that expedites and reduces the cost of delivering cash on the ground in crisis areas. Project Bifrost will serve as a critical lifeline to communities in a crisis to purchase food, medicine, clean drinking water, or cell phone minutes to connect with family and friends.[252]

GLOBAL ACCESS TO FINANCIAL INSTITUTIONS

According to the World Bank, 1.7 billion of the global population does not have access to financial institutions.[253] Many people think this lack of access only applies to developing nations, but financial institutions that provided profitable banking in the past have been leaving countries. According to the *Reuters.com* article, "Caribbean Counties Caught in the Crossfire of U.S. Crackdown on Illicit Money Flow," U.S. financial institutions have ended relationships with regional banks. For U.S. banks, the cost of providing financial services outweighs the risks as banks in the Caribbean are facing reduced profits and higher risks of money laundering.[254]

But there are two sides to this financial story. When enough banks leave a country, especially countries with small populations and economies, the local economy suffers. The local community loses an employer that provided quality jobs. Unless they are able to use online banking, many in the local population will be required to close their savings and checking accounts. Finally, remittances and foreign trade for businesses becomes more difficult. In short, life and business in the local economy becomes more difficult.

Cryptocurrencies provide alternatives for banking and remittances, both of which are critical to support local economies and provide affordable access to global markets. Banking institutions require government issued identification and minimum balances to open an account and transact business, which prevents billions of people from opening bank accounts. In some countries the minimum amount to open a bank account is $1.00 U.S., which may be a lot of money. Cryptocurrencies do not require the same level of requirements to open an account to transact business or remittances.

GLOBAL ACCESS TO FINANCIAL INSTITUTIONS FOR WOMEN

Cryptocurrency transactions and blockchain technology have a broad reaching impact on global diversity, especially in developing countries. Major global organizations consistently report women have been denied equal access to banking systems in their home countries. According to the World Economic Forum's Report, 1.1 billion women (55% of unbanked adults) are excluded from access to financial institutions. [255] The gender gap is particularly wide in South Asia where only 37% of women have a bank account compared to 55% of men. Similarly, the World Bank's Global Findex reports women in developing economies are 20%

less likely to have a bank account or they have accounts in a relative's name, most likely a male. [256] Cryptocurrency exchanges do not have any gender requirements, empowering women to control their own finances.

Anyone, regardless of gender, can open a cryptocurrency wallet or trading account and use the account as a means of financial freedom. This new financial freedom gives women the ability to: directly receive payment for employment, independently open and run their own business with complete control of the financial aspect of the company, and use cryptocurrencies as currency, investment, or blockchain business solutions. In some countries, these actions were not remotely possible until bitcoin was created.

RECORDING DEEDS ON BLOCKCHAINS

The comparison of the title search process to the blockchain process was covered in Chapter 5. This section will cover the positive real life uses of blockchain technology to benefit landowners that have not been able to use their existing land registry system to record the ownership of their land.

Blockchain land titling creates a transparent immutable process to record property ownership. The recording of land ownership is critically important since land is probably the owner's most valuable possession. In addition, especially in the case of farmers, the land may be the main source of income. Blockchain land titling protects the landowner from unjust claims against their land, but it also creates opportunities for the landowner to use the land as collateral once ownership has been established.

Blockchain land titling also serves communities impacted by natural disasters. For example, in 2010, a 7.0 earthquake devastated Haiti, resulting in the tragic loss of 230,000 lives. The international humanitarian response was immediate and started Haiti on a path to recovery. Eight years later, Haiti is in long-term recovery and dealing with a major issue: land titling. According to the *International Federation of the Red Cross and Red Crescent Societies (IFRC) Report*: "Impact of the Regulatory Barriers to Providing Emergency and Transitional Shelter After Disasters, Country Case Study: Haiti," one of Haiti's main recovery problems includes issues related to housing, land, and property. Haiti has weak regulations and has not been able to enforce property rights to the point that local authorities are not sure of ownership.[257] The lack of government accuracy to determine property ownership is sure to create mass confusion in any country.

A blockchain is a viable solution for Haiti and other problems with land titling. A blockchain titling system in Haiti could create a transparent permanent structured formal process rather than an "informal/customary" process to provide procedures to establish landownership. Haitian-Americans personally shared their concerns that real estate title problems have prevented the sale of land, resulted in multiple claims of ownership on the same parcel of land, and prevented much needed construction in many areas of Haiti.

Numerous articles report countries have already started to incorporate blockchain land titling. *Reuters.com*, "African Startups Bet on Blockchain to Tackle Land Fraud," tells troubling tales in Nairobi of cartels colluding with officials to create parallel land titles to illegally acquire land. Blockchain technology's transparent and immutable nature makes this illegal practice impossible to achieve.[258]

Another article in *Wallet Weekly* ("Blockchain Land Records: Six Countries That Are Testing the Technology As We Speak") reports the following countries are digitizing their land registries onto a blockchain:

1. Brazil
2. India
3. Russia
4. Sweden
5. Ukraine
6. United Kingdom

Russia, Sweden, Ukraine, and the United Kingdom are testing methods to use blockchain in land registries on a national scale, while localities in Brazil and India are leading the blockchain titling efforts in their country.[259] Similar to the blockchain efforts in Brazil and India, blockchain supporters in smaller countries advocate testing blockchain real estate titling in their countries. For example, the *Medium* article, "Use Blockchain for Land Titling in Jamaica," points out the benefits of blockchain land titling in Jamaica:

> "Jamaica is a small country and it would be relatively easy to put in place the things needed to eventually use blockchain to facilitate [the] land titling process, thereby unlocking the massive dormant value and helping to uplift people out of their current state. Imagine a small farmer now finally having undisputed proof of land ownership and being able to secure a low-interest loan from a bank to purchase the tractor and supplies to greatly improve the yields from the land? Now multiply that across the entire country and we can see how technology can directly be applied to help Jamaica achieve real economic growth that is inclusive, broad based, and helps us to leap-frog many developed countries."[260]

Landowners who purchased or inherited land deserve to use the land to fulfill their basic necessities without the threat of losing it to crime or corruption. A blockchain can provide solutions to these land-titling problems.

BLOCKCHAIN PROVIDES ENERGY IN NATURAL DISASTERS

Tragically, many natural disasters result in loss of life, but the loss of life may not end when the disaster is over. If the natural disaster damaged any part of the power transmission process, loss of power can contribute to loss of life in its aftermath. Loss of power can result in: loss of critical food sources, loss in communications to notify the authorities of personal emergencies and location for assistance, disruption of medical services that require power to operate, storage of medications requiring refrigeration to prevent expiration, and many other critical necessities. This section focuses on blockchain technology's impact on energy production during a natural disaster.

In a natural disaster resulting in power loss, the local power company is the only entity capable of restoring power. In some areas, even in the U.S., the local power company may not be able to restore power before having a lasting impact. For example, in 2017 two category five hurricanes, Irma and Maria, slammed the United States Virgin Islands and Puerto Rico. It took days, even weeks, for the local power companies to restore power to the major population centers of the islands.

According to an August 7, 2018, *CNN* article, it took eleven months for Puerto Rico Electric and Power Authority (PREPA) to restore power to nearly 1.4 million customers, leaving only 25 customers without power. [261] How does blockchain technology solve this daunting problem? One blockchain energy solution is a blockchain powered microgrid.

In the blockchain microgrid, houses and businesses support a device that collects and stores energy from natural sources (solar, wind, etc.). The stored energy is recorded on a blockchain microgrid and distributed to other registered users on that blockchain. The Brooklyn Microgrid, located in Brooklyn NY, is an example of a blockchain microgrid.

Brooklyn Microgrid users share power with the other microgrid users but they also have a partnership with a local power company. The mission of the Brooklyn Microgrid is to:

- Increase the amount of clean, renewable energy generated locally.
- Develop a connected network of distributed energy resources, which improves electrical grid resiliency and efficiency.
- Create financial incentives and business models that encourage community investment in local, renewal energy.[262]

So far, the Brooklyn Microgrid is accomplishing their mission statement objectives. As a local blockchain microgrid, all three objectives are critically important and should be used as a model for future blockchain microgrids. In the case of a disaster when the local power grid is down, Brooklyn Microgrid users have access power when they need it the most.

As more blockchain energy microgrids establish themselves, local power companies may have to redesign their profit centers to provide more support to microgrids in addition to providing power to local customers. This approach will also reduce local power company expenditures on installing and maintaining green energy equipment as the local blockchain power grids will invest in green energy equipment for their blockchain power grids.

BLOCKCHAIN AND THE ENERGY CRISIS IN DEVELOPING NATIONS

Accessible and reliable energy is an essential element for prosperity. For example, electricity powers sanitation systems, medical systems, and many food storage and transport systems. Unfortunately, 1.2 billion people in many developing areas of the world do not have access to sustainable energy. Many rural areas do not have power because they are geographically distant from sources of power generation and the cost of constructing and maintaining power lines may be too expensive to deliver power to rural areas.[263] Enter blockchain technology.

Blockchain technology gives rural areas the ability to create a microgrid to share solar, wind, or other power harnessed outside of the local power grid. ImpactPPA is an example of a blockchain company using blockchains to provide reusable electricity to rural areas. ImpactPPA summarizes their model to deliver power on a global scale:

> "With utility scale or Micro-Grids deployed in over thirty-five countries, power is generated, stored, and delivered. Power flows to a smart meter, which is connected to a blockchain. Consumers of electricity interact with their smart meter and purchase power from a mobile device."[264]

In addition, ImpactPPA's white paper abstract summarizes the energy problem and the blockchain solution:

> "Energy is the key to improving quality of life, yet approximately 1.2 billion people across the globe lack access to clean, reliable electricity. Distributed, renewable energy solutions empower underserved and impoverished communities—both literally and figuratively—while they reduce the use of fossil fuels and mitigate the effects of climate change. Many econo-

mists agree that in the coming years, great wealth creation will emerge from these 1.2 billion people who will ascend to middle class status. But ImpactPPA believes that this can only happen if they are given access to energy."[265]

ImpactPPA's decentralized blockchain energy platform allows users to prepay for the use of electricity with a smart phone app. ImpactPPA customers do not need a bank account for access to electricity. Through blockchains, ImpactPPA is bringing reliable and accessible power to the populations that need it the most. The next section will discuss blockchain technology's impact on access to clean water, another critical necessity.

CRYPTOCURRENCIES AND THE WATER PROJECT

Most of the developing world takes clean water for granted, but the Water Project understands the critical importance of clean water as well as the fundraising potential of cryptocurrencies. The Water Project was one of the first charities to accept cryptocurrencies. According to TheWaterProject.org, over 100 reliable water projects have been funded by cryptocurrency donations. TheWaterProject.org has over 209 stories of how clean water has improved the lives of people in the community. The impact of three clean water stories from The Water Project Kenya follow:

> Imbiakalo Community: "Before, our children used to have running stomach because of drinking water, and typhoid was also a great threat among elderly people who spent most of their time in hospital beds seeking treatment. Instead of spending time in hospitals, we now spend time on our small farms planting and harvesting crops."[266]

Kwambiha Community: "In the past, we used to lose more lives, especially young children, due to the outbreak of diarrhea and typhoid diseases. We no longer experience that because we now have access to clean and safe water within our reach."[267]

Kiluta Sand Dam: "The group members have successfully planted trees on their farms. Some of which have already started bearing fruits and are earning an income for the groups."[268]

These stories put the use of cryptocurrencies and the blockchain into perspective. The health of children and the elderly as well as the ability to grow crops are critical to the future of any community. Cryptocurrencies added a new dimension to the lifesaving work of The Water Project.

The next section will discuss the impact of cryptocurrencies and the blockchain on an Olympic level.

THE JAMAICAN BOBSLED STORY

In the 1993 hit movie *Cool Runnings*, the Jamaican Bobsled team competed in the four-man bobsled event in the 1988 Calgary Winter Olympics. The Jamaican Bobsled team lost control of the sled and crashed during the qualifiers, but as the ultimate underdogs, they won the hearts of the world. Let's do a "crypto sequel" of the Jamaican bobsled story. According to *The Guardian*, in 2014, a group of supporters raised $25,000 in Dogecoin, a decentralized peer-to-peer digital currency that enables you to easily send money online. This enabled the Jamaican two-man bobsled team to attend the Winter Olympics in Sochi, Japan.[269] Dogecoin's "Olympic" fundraising effort is another example of the positive real life impact of cryptocurrencies.

THE UNITED NATIONS' USE OF BLOCKCHAINS

According to the *GBA Global Newsletter*, The United Nations World Food Program is using digital vouchers in a blockchain pilot program to distribute $1.3 billion in cash entitlements to Syrian refugees. The pilot program includes approximately 100,000 Syrian refugees and is saving the program approximately $40,000 per month. When the program is expanded to include the full $1.5 billion in payments, the anticipated cost savings will be $1.6 million per month. That savings means more food will feed hungry and displaced Syrian refugee families.[270]

The United Nations (UN) Mission created a UN Blockchain, a Multi-Agency UN Platform. The UN Blockchain website creates a vision for the future of the world when blockchain technology is used to solve global problems.

The UN Chronicle, is the magazine of the United Nations, and one article in particular, "Blockchain and Sustainable Growth,"[271] talks about the theoretical, global use of blockchains by covering many of the main components in this book including: the various blockchain applications, the critical need to educate everyone on blockchain technology, and the importance of blockchain education to name a few. But one thing is clear from the article, blockchain technology is a global game changer, and the "game" is just starting. It will be decades before the true impact of blockchain technology is felt throughout the entire world, but the world will feel the change.

Although it is not specifically mentioned in the article, I personally believe global organizations like the United Nations will play a pivotal role in global blockchain education and implementation.

The final words of this article provides an excellent way to end this book:

"[T]he concept of citizen-held and citizen-owned solutions to global problems has been unleashed. The established international system ignores that message at its peril."[272]

Or in other words, *Blockchain or Die.*

THE GLOBAL BLOCKCHAIN FUTURE

On many occasions, I have been asked, "Eric, what is the future of the blockchain?" In one way, that is an easy question to answer: "Blockchains will change everything. They will change life as we know it." But the last words of that sentence, "life as we know it," is the mysterious key to the second answer to the same question.

No one exactly knows how blockchains will change the world in the next twenty years. In 1990, no one could predict Facebook, Twitter, Uber, Airbnb, eBay, Tinder, PayPal, Snapchat, and WhatsApp; but these internet-based technologies changed our lives forever.

That being said, I can make two general predictions, which have been discussed in *Blockchain or Die* numerous times: 1) Blockchain technology's decentralized peer-to-peer nature will eliminate many intermediaries. It's only a matter of time. Intermediaries need to find a way to use blockchain technology or face business extinction. In other words, "blockchain or die." 2) Governments on the local and state level are researching—and even using—blockchain technology in their governance models. National governments will have to decide what role they want cryptocurrencies and blockchain technology to play in their governance models. Governments had to learn how to use the internet and now almost everything is online; therefore, governments will have to learn to use blockchains as well.

SUMMARY

This chapter presented examples of the real life impact of cryptocurrencies and blockchain technology. When a technology directly or indirectly improves access to one of the most basic needs of every human being; that is powerful technology indeed. There are many more stories, and they will only increase in number as blockchain technology expands.

Thirty years ago, most people were asking, "what is the internet?" and now everyone in every part of the world uses the internet every day. In fact, we have become so dependent on the internet, it's difficult to imagine conducting business, paying bills, or interacting with people without it.

In the present day, most people are asking the question, "what is the blockchain?" Thirty years from now, people will be as reliant on blockchains as they are on the internet. Just like my Kenyan friend stated: "M-Pesa is life"; to many people, blockchain technology will give them new opportunities in life.

There are many challenges ahead for this new and wonderful technology and the primary challenges to global adoption includes: the role and impact of legislation, misunderstanding the technology, unethical businesses that will use the new technology to defraud people and investors, the threat to current centers of power and control, and the financial freedom that will loosen the grip of greed from those who prey on the unrepresented masses. That may sound idealistic, but remember, five countries passed legislation making any use of cryptocurrencies illegal. These challenges are real, but history has shown us that there is one thing that can defeat the centers of power that do not represent the will of the people, and that is the will of the people when they are focused and empowered to accomplish one goal. In this scenario, blockchain technology provides the focus and empowerment and the goal is financial independence. A powerful goal indeed.

SUCCESS STRATEGY ACTION ITEMS

1. If you send or receive remittances through the standard remittance channels, you are either paying too much money as a sender or receiving too little money as a receiver. Consider using a cryptocurrency platform for remittances, especially if sent internationally. Research the best cryptocurrency for your remittance needs and start by remitting a small amount of money as a test. Once both parties are comfortable with the cryptocurrency remittance process, the transactions can increase from the smaller test funds to the regular remittance amount.

2. A lot of programs are starting to use blockchain technology to bring more relief and aid to those in need. Please consider donating to these organizations.

3. If you are interested in the humanitarian impact of cryptocurrencies and blockchain technology, the United Nations website is an excellent resource to learn more about blockchain humanitarian efforts.

4. Determine how blockchain technology can help your community. The Brooklyn Microgrid is a prime example of a coordinated effort that combines businesses, utilities, and residents into a successful blockchain project.

AFTERWORD

Congratulations on completing *Blockchain or Die*. Cryptocurrencies and blockchain technology are new and exciting concepts, but since it is so new, it is also very challenging. Although it is challenging, it is the money and business of the future.

Now the question is, "what are you going to do with the information you learned?"

Blockchain or Die is not designed to make you an expert in cryptocurrencies and blockchain technology, but it is designed to give you tools to start your cryptocurrency and blockchain technology journey.

Eric L. Guthrie

Eric L. Guthrie, Esq. CDE

Appendix

CRYPTOCURRENCY AND BLOCKCHAIN TECHNOLOGY RESOURCES

"Getting information off the internet is like taking a drink from a fire hydrant."

—Mitchell Kapor, American Entrepreneur and Creator of Firefox, an open source web browser

THIS QUOTE VERY COLORFULLY describes the amount of information on the internet related to cryptocurrencies and blockchain technology. One of the major goals of this book is to provide the basic information on cryptocurrencies and blockchain technology in one place. Another goal is to provide resources to further your research and education on both topics at your own pace.

As I was writing *Blockchain or Die*, I asked people who attended my speeches on cryptocurrencies and blockchain technology to provide feedback on the kind of information that would make this book an excellent resource for them and others who are new to the world of cryptocurrencies and blockchain technology. Fortunately, many of the recommendations were already covered in the existing content, but there was one recommendation that was mentioned often enough that it called for its own section: a list of cryptocurrency and blockchain technology resources.

This appendix fills the resource and information gap that many people have expressed when discussing their challenges in learning about cryptocurrencies and blockchain technology. Review as much of the information provided as you can before investing in cryptocurrencies or before engaging in blockchain technology for your company.

CRITERIA FOR SELECTION AS A RESOURCE

Every resource in this section had to meet specific requirements for inclusion. It is important to understand that criteria as presented below:

- The resource must be available in English, but this does not preclude availability in other languages.
- Except for government websites, the resource must be specifically dedicated to cryptocurrencies or blockchain technology. Resources that are not specifically dedicated to cryptocurrencies or blockchain technology were not included. For example, a website or blog that discusses general investment issues or business issues and contains occasional information on cryptocurrencies or blockchain technology was not included.
- Must have been accessible prior to the submission of *Blockchain or Die* to the editing process.
- Most of the resource organizations are based in the United States, but the list is inclusive of international organizational resources.

Also, this list does not represent all of the available resources on this topic. There are tens of thousands of resources worldwide on this technology. If you know of a competent reliable resource, you are encouraged to continue to use your resource.

This appendix presents the resources in three general categories: joint resources, cryptocurrency resources, and blockchain technology resources. Note that if a web address exceeds the length of a line, it will wrap onto the next line with a hanging indent and will *not* be hyphenated.

JOINT RESOURCES

UNITED STATES GOVERNMENT WEBSITES

The following resources have been compiled from U.S. government websites. While many local, state, and international government websites contain information on cryptocurrencies and blockchain technology, there are too many to list in this appendix. You are encouraged to visit government websites in your area to identify government resources.

Securities and Exchange Commission
https://www.sec.gov/news/public-statement/statement-clayton-2017-12-11
https://www.investor.gov/additional-resources/specialized-resources/spotlight-initial-coin-offerings-digital-assets

Commodities and Futures Exchange Commission
www.cftc.gov

Internal Revenue Services
https://www.irs.gov/newsroom/irs-virtual-currency-guidance
https://www.irs.gov/businesses/small-businesses-self-employed/virtual-currencies

Federal Trade Commission
https://www.consumer.ftc.gov/blog/2018/02/know-risks-you-invest-cryptocurrencies

Department of Homeland Security
https://www.dhs.gov/CSD-ANC

National Institute of Science and Technology
https://csrc.nist.gov/CSRC/media/Publications/nistir/8202/draft/documents/nistir8202-draft.pdf

United States Congress
https://www.congress.gov/115/crpt/hrpt596/CRPT-115hrpt596.pdf

General Services Agency
https://emerging.digital.gov/blockchain-resources/

WHITE PAPERS

White papers are best way to learn as much as you can about a specific cryptocurrency. White papers are written by the creator(s) of the cryptocurrency so you are getting the information directly from the source. The creators and executives make the white paper accessible as it is the primary document investors use to determine if they are interested in investing. The white paper is usually on the website home page of the cryptocurrency. In addition, white papers can be found on the following sites:

www.WhitepaperDatebase.com
www.StartupManagement.org

BOOKS

There are many published books on cryptocurrencies and blockchain technology. As this is a new technology, many of the books are introductory books or tell the history of this new technology.

Blockchain by Melanie Swan
Blockchain: Discover the Technology behind Smart Contracts, Wallets, Mining and Cryptocurrency (including bitcoin, ethereum, ripple, digibyte and others) by Abraham K White
The Age of Cryptocurrency: How Bitcoin and the Blockchain Are Challenging the Global Economic Order by Paul Vigna and Michael J. Casey

Blockchain Revolution: How the Technology Behind Bitcoin Is Changing Money, Business, and the World by Don and Alex Tapscott

Blockchain Technology Explained: The Ultimate Beginner's Guide About Blockchain Wallet, Mining, Bitcoin, Ethereum, Litecoin, Zcash, Monero, Ripple, Dash, IOTA and Smart Contracts by Alan T. Norman

Blockchain Basics: A Non-Technical Introduction in 25 Steps by Daniel Drescher

The Internet of Money by Andreas M. Antonopoulos

The Book of Satoshi by Phil Champagne

Digital Gold: Bitcoin and the Inside Story of the Misfits and Millionaires Trying to Reinvent Money by Nathaniel Popper

Mastering Bitcoin for Dummies: Bitcoin and Cryptocurrency Technologies, Mining, Investing and Trading—Bitcoin Book 1, Blockchain, Wallet, Business by Alan T. Norman

Cryptocurrency Investing Bible: The Ultimate Guide about Blockchain, Mining, Trading, ICO, Ethereum Platform, Exchanges, Top Cryptocurrencies for Investing and Perfect Strategies to Make Money by Alan T. Norman

The Age of Cryptocurrency: How Bitcoin and the Blockchain Are Challenging the Global Economic Order by Paul Vigna and Michael J. Casey

Mastering Bitcoin: Unlocking Digital Cryptocurrencies by Andreas M. Antonopoulos

The Science of the Blockchain by Roger Wattenhofer

Blockchain: The Beginners Guide to Understanding the Technology Behind Bitcoin & Cryptocurrency (The Future of Money) by Artemis Caro

The Business Blockchain: Promise, Practice, and Application of the Next Internet Technology by William Mougayar

Bitcoin and Cryptocurrency Technologies: A Comprehensive Introduction by Arvind Narayanan, Joseph Bonneau, Edward Felten, Andrew Miller, and Steven Goldfeder

Blockchain Basics: A Non-Technical Introduction in 25 Steps by Daniel
    Drescher
Blockchain: Down the Rabbit Hole by Tim Lea
Bitcoin The Future of Money by Dominic Frisby
Wildcat Currency by Edward Castronova
The Bitcoin Bible by Benjamin Guttmann

MESSAGING APPS

Messaging apps are social messaging platforms that allow the messaging platform to communicate messages to their subscribers.

Telegram—[Every ICO has a telegram, nonspecific telegrams tend to
    revolve around trading]
http://t.me/TheCryptoBible
http://t.me/cointelegraph
https://t.me/bitcoinBravado
https://t.me/Crypto
http://t.me/UpcomingICOs
http://t.me/blockchaininvestio
http://t.me/bitcoinchannel
https://t.me/cryptocurrency
http://t.me/cryptoTalkC
https://t.me/CryptoCurrencies
https://t.me/CryptoCommunities
https://t.me/bitafta Cryptocurrency Technical Analysis
http://t.me/CryptoJobs
http://t.me/bitcoinChat
http://t.me/preicoved Pre-ICO and Pre-sale crypto projects
https://t.me/join_cht CryptoHackers Community
http://t.me/cryptoaquarium
https://t.me/bitcoinCore
https://telegram.me/wallstreetTraderSchool
http://t.me/CoinDesk Blockchain and Crypto News (not official)

http://t.me/cointrendz
https://t.me/whalepoolbtcfeed
https://t.me/cryptodirectory
https://t.me/askmecryptoss
https://t.me/Cryptolearn Trading and Investment
https://t.me/bigorg Blockchain Interest Group

FORUMS

Forums play an important role by using the internet to create a platform learning, and sharing information, thoughts, and ideas with other members of the forum.

https://bitcointalk.org/
https://bitcoingarden.org/forum/
https://forum.bitcoin.com/
http://www.altcoinstalks.com/
http://bitco.in/forum/
http://www.bitcointrading.com/forum/index.php
http://coinboards.org/
https://blockchainforums.info/
https://thecryptocurrencyforums.com

REDDIT

Reddit calls itself "the front page of the internet." Reddit is a collection of forums where people can share news, information, and ideas on virtually everything, including cryptocurrencies.

https://www.reddit.com/r/CryptoCurrency/
https://www.reddit.com/r/ico/
https://www.reddit.com/r/CryptoCurrencies/
https://www.reddit.com/r/CryptoCurrencyTrading/
https://www.reddit.com/r/CryptoTechnology/

https://www.reddit.com/r/bitcoin/
https://www.reddit.com/r/altcoin/
https://www.reddit.com/r/crypto/
https://www.reddit.com/r/CryptoMarkets/
https://www.reddit.com/r/BlockChain/
https://www.reddit.com/r/BlockchainGame/
https://www.reddit.com/r/Blockchain_Healthcare/
https://www.reddit.com/r/Crypto_Currency_News/
https://www.reddit.com/r/bitcoinBeginners/
https://www.reddit.com/r/CryptoTrade
https://www.reddit.com/r/Jobs4bitcoins
https://www.reddit.com/r/bitcointechnology

DISCORD

Discord is a text and chat application. Originally Discord was created for gamers, but since gamers were one of the first users of bitcoin, it was commonly discussed in Discord.

https://blockchaindiscord
https://discord.me/blockchaindevs
https://discordlist.me/join/310180006653591555/—Blockchain
    Discord
https://discord.me/cryptochain
https://discord.me/hypedoncrypto
https://discord.me/etherealm
https://discord.me/instantcryptoalerts
https://discordlist.me/join/392600094471356417/ cryptotalk
https://discordapp.com/invite/XMTTPWK crypto discussion—Kripto
    Community
https://discordapp.com/invite/UeVMP9D "Largest active crypto
    discord" Spacestation
https://discord.gg/DdBeJBbCryptocurrency Book 5.3.18.docx
    cryptopedia

https://discord.gg/RbVNEaZCryptocurrency Book 5.3.18.docx hive
   cryptos

https://discord.gg/DpGXJEC Pip Guru (trading signals and analysis)

https://discord.gg/VgysgHXCryptocurrency Book 5.3.18.docx crypto
   mining

https://discord.gg/4gvfHVJCryptocurrency Book 5.3.18.docx trade
   read

https://discord.gg/YtYUhUHCryptocurrency Book 5.3.18.docx ICOGO
   (investors)

https://discord.gg/eBhaF7T cryptocentral (trading)

ORGANIZATIONS

Meetups have become an important way to meet other people
who are interested in cryptocurrencies and blockchain technology. In many ways, bitcoin started as a community as a way to improve the bitcoin protocols and spread the word about bitcoin.
Once you create or join a cryptocurrency or blockchain Meetup
group, you are now part of a group where you can share:
thoughts, experiences, contacts, business contacts, and much
more.

Also, you do not have to limit yourself to one Meetup group. If
you have more than one cryptocurrency or blockchain Meetup
group in your area, you should join as many as your time permits,
especially if you are new to cryptocurrencies or blockchain technology.

You will find that different Meetup groups have different
areas of focus in cryptocurrencies or blockchain technology. For
example, my first cryptocurrency Meetup group focused on
educating their attendees with presentations related to the
technology and business model behind bitcoin and Ethereum. My
second Meetup group, the Government Blockchain Association, a
member-based 501(c)(6) organization, focused on the many

aspects of blockchain technology, including educating memberships and the blockchain community on government contracting and private companies in the government contracting space.

One afternoon, I was sharing my Meetup experiences to a friend who was interested in cryptocurrencies and blockchain technology. During our discussion, she asked "what else are the Meetup groups going to do for me?" I was surprised to hear her ask that question and my response was the more active you are in the Meetup group, the greater the return of the time you invested.

At this stage in cryptocurrency and blockchain technology, a large part of the group efforts are based on grassroots organization. Meetup members, and potential members, should consider bringing their own skill sets and experience to the Meetup group. For example, I am a Certified Training Director, and the cryptocurrency and blockchain groups had great content, they just needed to incorporate training principles focused on training the adult learner on the subject matter. As many cryptocurrency adopters in the past volunteered their services to further the progression of bitcoin, I volunteered my training expertise.

CRYPTOCURRENCY RESOURCES

Websites\Blogs

Generally, websites are a good place to start for information. The following websites will help you to learn more about cryptocurrencies.

https://www.coindesk.com/
https://www.ccn.com/

https://cointelegraph.com/

https://news.bitcoin.com/

https://cryptocoinsnews.com

https://newsbtc.com

http://bitcoinist.com/

http://forklog.net/

https://medium.com/wearetheledger

http://blog.genesis-mining.com/

https://99bitcoins.com/

https://cryptocurrencyacademy.blogspot.com/

http://cryptomining-blog.com

https://coindesk.com

https://coinmarketcap.com

https://cryptoclarified.com/

http://vectorspace.ai/smart_token_baskets/smart-token-basket.html

https://www.cryptocoinsnews.com/

http://insidebitcoins.com/

https://www.newsbtc.com/

http://abitco.in/

https://bitcoin.stackexchange.com/

https://bitcoinmagazine.com/

http://bravenewcoin.com/

ORGANIZATIONS

This book has consistently referenced the importance of organizations as a resource to learn more about cryptocurrencies. The following organizations are a good place to start.

https://cryptoconsortium.org/ Cryptocurrency Certification
    Consortium
https://idacb.com/ International Decentralized Association of
    Cryptocurrency and Blockchain
https://cryptocurrency.how/bitcoin&CryptocurrencyBusinessDirectory

http://www.crypsa.org Cryptocurrency Standards Association
  CRYPSA
https://www.worldccorg.com/

PODCASTS

Invest Like the Best: Hash Power series |
  http://investorfieldguide.com/hashpower/
Coin Mastery |https://www.coinmastery.com/
Bitcoin Uncensored | https://itunes.apple.com/us/podcast/bitcoin-
  uncensored/id1046414365?mt=2
The Bad Crypto Podcast | https://itunes.apple.com/us/podcast/bad-
  crypto-podcast-bitcoin-blockchain-ethereum-
  altcoins/id1261133600?mt=2
Lets Talk Bitcoin | https://letstalkbitcoin.com/
Coin Talk | https://itunes.apple.com/us/podcast/coin-
  talk/id1332061471?mt=2
Crypto Radio | http://cryptoradio.io/
Breaking Banks | https://breakingbanks.com/
Bitcoin Knowledge Podcast | http://www.bitcoin.kn

NEWS AGGREGATORS

https://coinhooked.com/cryptocurrency-news/
http://bitcoinagile.com/
https://coinspectator.com/
https://www.geekwrapped.com/cryptocurrency

MAGAZINES

http://nxter.org/
https://ybitcoinmagazine.com/
https://cryptocoremedia.com/
http://icocrowd.com/
https://bitcoinnewsmagazine.com/
http://www.thecryptocurrencymagazine.com/

https://www.21cryptos.com/
http://cryptonewsmag.com/
http://cryptocurrency-magazine.com/
http://cryptocurrencymagazine.com
https://en.decentral.news/
https://cryptohackers.party/

INSTAGRAM PAGES

https://www.instagram.com/ico_list/
https://www.instagram.com/icopresale/
https://www.instagram.com/sharecrypto/News
https://www.instagram.com/cryptohubofficial/newsandmemes
https://www.instagram.com/cryptosharkk/NewsMemesMotivation
https://www.instagram.com/cryptoshi_hackamoto/NewsMemesandFun
https://www.instagram.com/cryptoworld1001/NewsandVideos
https://www.instagram.com/cryptotroll/humor
https://www.instagram.com/cryptocompass/tradingsignals
https://www.instagram.com/thecryptograph/DailyCryptoNews
https://www.instagram.com/altcoingenius/news
https://www.instagram.com/crypto.diary/memes
https://www.instagram.com/ethereumtony/influencer
https://instagram.com/bitcoindoc
https://instagram.com/bitcoinsberlinincubator
https://instagram.com/bitcoincenternyceducationcenter
https://instagram.com/bitcoiniacs
https://instagram.com/bitcoinprice

YOUTUBE CHANNELS

BuriedONE Cryptomining | https://www.youtube.com/user/BuriedOne
Makaveli's Blockchain Network |
    https://www.youtube.com/user/makavelithedon2006
B21 Block: Cryptocurrency and Blockchain Tutorials |
    https://www.youtube.com/channel/UC1y2yA6CTN_O866SkeNnfYg

Cryptocurrency Market |
  https://www.youtube.com/user/PrisonOrFreedom
Cryptocurrency Vlog | https://www.youtube.com/channel/UCMW2-
  dnYi6R4eSlsowsBq8g
24/7 Cryptocurrency News |
  https://www.youtube.com/channel/UCiJ2JHYbuxuZkWqybsSYoFQ
Cryptocurrency Investments |
  https://www.youtube.com/channel/UC8NnPYYinxhIPBcchjlN1bw
Faze Crypto |
  https://www.youtube.com/channel/UCQFhDCeJTc25IrJHLiFWpZ
  Q
Cryptocurrency Investing |
  https://www.youtube.com/channel/UCSXItn5Tru2QZTEMS8prUT
  g
Coin Mastery |
  https://www.youtube.com/channel/UC4nXWTjZqK4bv7feoRntSog
Crypto Kirby—Bitcoin and Cryptocurrency Trading |
  https://www.youtube.com/channel/UCRKK14cbb7HXlz2O-
  V64zLQ

FACEBOOK GROUPS

https://www.facebook.com/groups/cryptosnews/
https://www.facebook.com/groups/TheCryptocurrencyAcademy/
https://www.facebook.com/groups/1393515590756580/—Crypto
  Gurus
https://www.facebook.com/groups/cryptoland/
https://www.facebook.com/groups/246479302178538/—Crypto
  Warriors
https://www.facebook.com/groups/CryptoCurrencyCollectorsClubPubl
  icForum/
https://www.facebook.com/groups/1863296943682020/—That crypto
  hustle
https://www.facebook.com/groups/LitecoinGroup/—Altcoin Investing

https://www.facebook.com/groups/480785485606417/—
Cryptocurrency Investing

https://www.facebook.com/groups/303525260097501/—
Cryptocurrency Investing \ News and Discussion

https://www.facebook.com/groups/111154582843020/—Crypto Hunt
Group

https://www.facebook.com/groups/TheBitcoin/

https://www.facebook.com/groups/cryptocurrencycollectorsclub/

https://www.facebook.com/groups/cryptortrust/—Bitcoin and
Blockchain Fintech Group

https://www.facebook.com/groups/1847056422214050/ the crypto
watch

TWITTER PAGES

https://twitter.com/jerrybrito

https://twitter.com/coindesk

https://twitter.com/vitalikbuterin

https://twitter.com/DerinCag/

https://twitter.com/CarrascosaCris_

https://twitter.com/niccary

https://twitter.com/NeerajKA

https://twitter.com/aantonop Andreas Antonopoulos, "The Bitcoin
Scribe"

https://twitter.com/NickSzabo4 Nick Szabo, "Not Satoshi, I Swear"

https://twitter.com/gavinandresen former bitcoin core developer

https://twitter.com/SatoshiLite

https://twitter.com/brian_armstrong Coinbase CEO

https://twitter.com/starkness

https://twitter.com/twobitidiot

https://twitter.com/AmberBaldet

https://twitter.com/lopp

https://twitter.com/coindesk

https://twitter.com/WhalePanda

https://twitter.com/rogerkver

https://twitter.com/Excellion
https://twitter.com/prestonjbyrne
https://twitter.com/ErikVoorhees
https://twitter.com/Melt_Dem
https://twitter.com/TuurDemeester
https://twitter.com/gendal
https://twitter.com/niccary
https://twitter.com/cdixon
https://twitter.com/TimDraper
https://twitter.com/NicTrades
https://twitter.com/Fehrsam
https://twitter.com/bhorowitz
https://twitter.com/PipCzar
https://twitter.com/catheryne_n
https://twitter.com/barrysilbert
https://twitter.com/jimmysong
https://twitter.com/balajis

BLOCKCHAIN TECHNOLOGY RESOURCES

WEBSITES\BLOGS

https://www.myblockchainblog.com/
https://www.hoganlovells.com/en/blockchain-blog
https://blockchainatberkeley.blog/
https://www.ibm.com/blogs/blockchain/
https://www.technologyreview.com/newsletters/chain-letter/
https://www.notey.com/blogs/blockchain
https://blog.blockchain.com
https://www.research.ibm.com/blockchain/
https://www.futureblockchainsummit.com/
http://blockchaincanada.org/

PODCASTS

Unchained | http://unchainedpodcast.co/

Epicenter | https://epicenter.tv/

Blockchain Dynamics | http://www.blockchaindynamics.net/

Blockchain Curated | https://www.blockchaincurated.com/

Block Zero | http://www.blockzero.show/

https://thebitcoinpodcast.com/

The Blockchain Show | http://www.theblockchainshow.com/

Software Engineering Daily—Blockchain |
    https://itunes.apple.com/us/podcast/blockchain-software-
    engineering-daily/id1230807219?mt=2

Blockchain Innovation | http://blockchain.global/blockchain-
    innovation/

The Third Web | https://itunes.apple.com/us/podcast/the-third-
    web/id899090462?mt=2

Explain Blockchain Podcast | https://explainblockchain.io/

YOUTUBE CHANNELS

Blockchain at Berkley |
    https://www.youtube.com/channel/UC5sgoRfoSp3jeX4DEqKLwKg

New Kids on the Blockchain |
    https://www.youtube.com/channel/UCrLXGsnSCgH0bOOtP32S2g
    w

Media Box Ent Blockchain |
    https://www.youtube.com/channel/UCthP1zMBFq34-n02iSLC8Hg

Blockchain Labs |
    https://www.youtube.com/channel/UCbnvueL1Amt6qkCXQBYexyg

Blockchain Conferences |
    https://www.youtube.com/user/BitcoinConference

Blockchain University |
    https://www.youtube.com/channel/UCJ5uHx90mZGlK0lC-GSmtzw

Blockchain Workshops |
    https://www.youtube.com/channel/UC9Lmf5FfNkSmYMoxhQh5kt
    A

Blockchain News | https://www.youtube.com/channel/UClY8-
  XHO88IeMpaJgR94w-w
Blockchain WTF |
  https://www.youtub8e.com/channel/UCKENI1wZKvZFjtgWGOq2p
  CQ
ICO Blockchain |
  https://www.youtube.com/channel/UCk1Y1XGpP2y4jPH6Wb8Yph
  g
Blockchain Wealth |
  https://www.youtube.com/channel/UCs2jNdnL9ImergCJ8p8iFWw

LINKEDIN LINKS

Cryptocurrencies, Ethereum and Blockchain—
  https://www.linkedin.com/groups/13555328/profile
Bitcoin Blockchain & FinTech Think Tank > Cryptor Trust
  https://www.linkedin.com/groups/6580131/profile
Blockchain Business—
  https://www.linkedin.com/groups/4458579/profile
Blockchain Tech—Block Chain Crypto ICO Distributed Ledger Smart
  Contract Solidity Ethereum Bitcoin -
  https://www.linkedin.com/groups/8446680/profile
Blockchain For Finance Professionals -
  https://www.linkedin.com/groups/13552895/profile
Blockchain Professional Network—
  https://www.linkedin.com/groups?gid=3763996
Bitcoin—Blockchain Network—
  https://www.linkedin.com/groups?gid=4243277
Bitcoin—https://www.linkedin.com/groups/3721050/profile
Blockchain and Bitcoin Startups—
  https://www.linkedin.com/groups/5190684/profile
Crypto World—https://www.linkedin.com/groups/956157/profile
Blockchain for Sustainable Supply Chains—
  https://www.linkedin.com/groups/13523369/profile

ORGANIZATIONS

https://www.gbbcouncil.org/—Global Blockchain Business Council
http://www.unwbo.org/—World Blockchain Organization
http://www.blockchains.world/—World of Blockchains
https://worldblockchain.org/—World Blockchain Association
https://www.blockchainassoc.org/—International Blockchain
    Association
http://blockchainalliance.org/
https://www.blockchainresearchinstitute.org/
http://www.untitled-inc.com/
https://www.wsba.co/
https://bitcoinfoundation.org/
http://worldblockchain.foundation/
http://nakamotoinstitute.org/
https://iobf.co/—Internet of Blockchain Foundation
https://www.theblockchainacademy.com/
http://philanthropyblockchain.org/—Blockchain Philanthropy
    Organization
https://bitcoincryptocurrency.com/ Blockchain News, Tokens, Guides &
    Culture
Distributed Ledger Foundation

MAGAZINES

https://www.blockchainmagazine.net/
https://bitcoinmagazine.com/
https://bittpress.com/
http://forklog.net/ Blockchain and Bitcoin
https://medium.com/zapchain-magazine
http://ledgerjournal.org/—Ledger is a peer-reviewed scholarly journal
    that publishes full-length original research articles on the subjects
    of cryptocurrency and blockchain technology.
https://btcmedia.org/
https://en.decentral.news/—Blockchain and Crypto web magazine
https://bittmint.com/ Blockchain News Mag

https://tokeniest.com/ Blockchain Magazine News

SUMMARY

This appendix contains a great deal of resources for research and review so you will not have to conduct your own. While this is still a "fire hydrant" of information, I hope that categorizing it and putting all in one place will give you a sense of control.

SUCCESS STRATEGY ACTION ITEMS

1. Everyone processes information in different ways. The resources in this appendix have video, audio, and reading materials. Choose the method of learning that suits you best.

2. Set reasonable research expectations. Pace yourself. Don't try to process all of this information all at once. One recommendation is to read a few articles a day and don't move on to the next set of articles until you fully understand the articles you just completed.

3. If you have not already done so, set up your social media accounts (Facebook, LinkedIn, Instagram, Twitter, or Reddit) and set up messaging apps (telegram) to receive information and updates.

4. Although these resources are extremely informative, it is very helpful to talk to fellow blockchain enthusiasts. Conversations with fellow blockchain enthusiasts are excellent ways to learn more about the blockchain. These will help you process the massive amounts of information and give you a different perspective.

GLOSSARY

This glossary only covers the terms used in the book, but in most cases it provides a deeper explanation. There are many other cryptocurrency and blockchain definitions you should learn as you continue your *Blockchain or Die* journey.

Address/Public Address—A string of 26-35 alphanumeric characters, beginning with the number 1 or 3, that identifies a cryptocurrency wallet. It is used as a way to safely receive cryptocurrency. www.Blockchain.com gives a sample bitcoin address: 1BoatSLRHtKNngkdXEeobR76b53LETtpyT

Altcoin (or "alternate coin")—Any cryptocurrency except for bitcoin. "Altcoin" is a combination of two words: "alternative bitcoin" or "alternative coin." Examples include: Ripple, Steem, Monero, etc.

ASIC/ASIC Miner (Commonly known as an application-specific integrated circuit or ASIC)— "Integrated circuit" is just a computer chip. "Application-specific" means it was built for one specific purpose or computer application rather than a general-purpose application. An ASIC is used in cryptocurrencies to help record transactions on the digital record or blockchain. This process is known as mining. In bitcoin mining hardware, ASICs were the next step of development after CPUs, GPUs, and FPGAs. ASICs are capable of easily outperforming the aforementioned platforms for bitcoin mining in both speed and efficiency. Note that bitcoin ASIC chips generally can only be used for bitcoin mining.

Block—A single digital record created within a blockchain. Each block contains a record of the previous block, and when linked together these become the "chain."

Blockchain—A digital ledger, used to prove a group of people came to an agreement about something. By using this ledger every user is able to find out what amount of bitcoin has ever belonged to a particular address at a certain time period. The blockchain is supported by decentralized efforts of many miners. With this information, anyone can find out how much value belonged to each address at any point in history. The blockchain is very unique because it is simultaneously created and maintained by thousands of individuals. Blockchain records are permanent and very secure, preventing manipulation. Imagine the blockchain as a book of records. Every page in the book is a block in the blockchain and every page records any type of information. Blocks are created one after the other and chained together in sequence. Updates are often seen in minutes or even seconds and manipulation is extremely difficult, nearly impossible.

Decentralized Application (DApp)—A software application that has its technology running publicly on a network of computers. DApps are maintained by many individuals instead of by one organization. That network gives the technology security. A hacker cannot alter the application's data unless they were able to get access into nearly all of the network's computers and adjust it there.

Decentralized Autonomous Organizations (DAO)—A leaderless organization supported by a network of computers. To be decentralized, it must have no central location because it is running on a network of computers. And because there is no single leader and has its own rules to follow, it is autonomous, or self-governing.

Digital Signature—Permission and proof done through a computer that an authorized person has agreed to something and generates a verification code that proves a transaction took place. Digital signatures are used by cryptocurrency systems to allow the owner to send and receive money. When a signer authorizes something, they use their private key known only to them, to encrypt information along with a stamp of the time of signing. If the information is somehow modified, the time stamp will be altered and the digital signature becomes void and invalid. The person receiving or verifying the signed and encrypted information uses the signer's public key to verify the information came from the signer.

Double Spending—A form of deceit using digital money where the same money is promised to two parties but only delivered to one. If completed successfully, one of the two recipients will receive worthless money. Double spending may sound impossible, but digital technology enables the copying of anything in a digital format, including digital currency. Because the blockchain is public, many people are simultaneously verifying and recording information on it using their computers. After enough users in this network confirm your transaction, the guy who wants to double spend cannot.

Electronic Cash/Cryptocurrency—Electronic money that uses technology to control how and when it is created and lets users directly exchange it between themselves, similar to cash. For example, the proof-of-work protection method and asymmetric encryption, and the system functions decentrally in a distributed computer network.

Fork/Hard Fork—A decision to make a permanent change to the technology used by a cryptocurrency. This change makes all new recordings (blocks) very different from the original blocks. They are changed so much that new blocks are seen as invalid to anyone who didn't upgrade their technology. Which means, any computer that is not updated with the new technology, will find these new blocks appear invalid. Any alteration to bitcoin changes the block structure (including block hash), difficulty rules, or increases the set of valid transactions is a hard fork.

Fork/Soft Fork—A change made to cryptocurrency technology creating a temporary split in the group of recordings (blockchain). This change creates all new, valid recordings (blocks) that are slightly different from the original blocks. They are just different enough that users of the new technology see blocks from original technology as invalid. But, users of the original technology see no problem with either one.

As a result this means, new blocks will work just fine for all computers including those using the original technology. But computers using the original technology will find their blocks are rejected by the rest of the network until they upgrade and rejoin the network. Any new fork in the blockchain can fail and if it does, all users will return to the original recording. A change to the bitcoin protocol wherein only previously valid blocks/transactions are made invalid.

Forming a Record/Ledger—A book or other collection of records in which a person, business, or other group records how much money it receives and spends. Although the word "ledger" is not actually used in the abstract, the phrase "forming a record" is used and is an extremely important reference serves as the basis for many of the definitions that follow. The "record," which is officially called a "ledger," is the product that makes bitcoin, and many cryptocurrencies, possible as a decentralized and deregulated currency. That may not sound critical, but the next phase, as stated in the white paper, shows the importance of the ledger since it "cannot be changed without redoing the proof of work."

Gas—A small amount of Ethereum paid to people who use their computers to record transactions and do other software actions. Gas is calculated by multiplying a very small amount of Ethereum, known as gas price or gwei, and multiplying that by how much gwei you want to spend known as gas limit. Because 1 Ethereum = 1 billion (1,000,000,000) gwei, gas costs are usually very small, around several dollars. If the amount of gas is insufficient to complete the work, the work will fail. On the other hand, you can pay a bit more gas and expect the computers to complete your task sooner.

Hash—A computer program that takes information and irreversibly turns it into a series of letters and numbers of a certain length. Like all computer data, hashes are large numbers, and are usually written as a hexadecimal. The same hash will always result from the same data, but modifying the data by even one bit will completely change the hash. Bitcoin uses the SHA-256 hash algorithm (secure hash algorithm) to generate verifiably "random" numbers in a way that requires a predictable amount of

CPU effort. If you want to see the SHA-256 algorithm, https://www.movable-type.co.uk/scripts/sha256.html has a SHA-256 hash "converter" that converts any message into a SHA-256 hash. For example, the SHA-256 hash message of "Eric Guthrie" is

e461d5bba6eb11b6372d5ba9c54fe882f5d6f170483411018813a 7c4225743d2 in the SHA-256 hash.

Hashing—The actual work done by the central processing unit (CPU) to confirm the electronic transactions (bitcoin transactions). Consider hashing like a transparent electronic audit where all transactions have to be confirmed by the peer-to-peer network before the transaction can be finalized.

Initial Coin Offering (ICO). Also "Initial Token Offering" or "ITO" and "Token Generation Event" or "TGE")—When a new cryptocurrency or token generally becomes available for public investment. ICOs are similar to Initial Public Offerings (IPOs) where a company raises money by selling public shares of their stock. An ICO can be made for cryptocurrencies that are still in the idea stage, or are already built and ready for distribution. People invest their money in an ICO hoping they will greatly increase in value. For example, Ethereum sold their coins in their 2014 ICO for $0.40 and in 2017 the coins were worth over $400 each.

Mining—The process of using computer power to solve a complex math problem presented by the crypto system, review and verify information, and create a new recording to be added to the blockchain. In mining, transactions are added to the transaction records to public ledger of past transactions. This ledger of past transactions is called the blockchain, as it is a chain of blocks. The blockchain serves to confirm transactions to the rest of the network as having taken place.

The first miner to solve the problem creates a new block and receives compensation. To keep the blockchain network running smoothly, only one block can be created at a time. Proof of work is the mining process of controlling how blocks are created and how data is added to a block.

Nodes—Any computing device (computer, phone, etc.) that is participating in a network by way of receiving and sending data. Any computer, a phone, or any other computing device that can receive, transmit, and/or contribute to the blockchain is a node. Nodes that fully enforce all of the rules of bitcoin are called full nodes. Consider a node as a "bitcoin bee" working with thousands of other bees to support the function of the bitcoin hive. Bitcoin nodes use the blockchain to distinguish legitimate bitcoin transactions from attempts to re-spend coins that have already been spent elsewhere. Without nodes, bitcoin transactions would not be possible.

Peer to Peer /P2P—A connection between two or more computers that allows them to directly share information, files, or other data. In a peer-to-peer network, there are no privileged peers with special rights or advantages. Every peer has the same rights and they all share the workload. Another way to define "peer-to-peer" is to start with the rest of the sentence in the Bitcoin White Paper: "one party to another." "One party to another" means two parties that interact using a network to transact business without going through a centralized system, such as a financial institution. Peer-to-peer cryptocurrencies are designed to decentralize the transfer of money, or "electronic cash."

Private Key—A string of random letters and numbers known only by the owner that allows them to spend their cryptocurrency. The private key is mathematically related to the cryptocurrency address, and is designed so the cryptocurrency address can be calculated from the private key, but importantly, the same cannot be done in reverse.

Never share your private key unless you want someone else to have access your money!

Your private key is very similar to your password to access your crypto. Compare a private key with a public key and address:

- Your public key is rarely ever used, but you can use it to receive cryptocurrency.
- Your address is a safer version of your public key and is what you should use to receive money.

Proof of Work—A process for achieving consensus and building on a digital record on the blockchain. In Proof of Work, computers compete to solve a tough math problem. The first computer, defined later as a "node" to solve the problem is allowed to create new blocks and record information earning them a reward in digital currency plus fees paid for each transaction. Producing a proof of work can be a random process with low probability so that a lot of trial and error is required *on average* before a valid proof of work is generated.

Public Key—A code consisting of a string of letters and numbers that allows cryptocurrency to be received. Public keys are not considered as safe to use as public addresses. A "public key," it isn't publicly visible until you've shared it or sent money out. Every bitcoin address has a public key, which coupled with the private key, ensures the security of the crypto-economy.

Smart Contract (also self-executing contract, block-chain contract, or digital contract)—An agreement to exchange goods, services, or money that will automatically execute, without third party oversight, so long as established criteria are met. Smart contracts are electronic algorithms that automate the contract execution process in the blockchain. Smart contracts standardize, automate, and exclude divergences in the treatment of the agreement terms by the entered parties. Smart contracts have the ability to receive, store, and send cryptocurrencies. Once they receive money, they will automatically do whatever they were programmed to do like sending money. When the smart contract has completed its task, it shuts down.

Smart Property—Property whose ownership is controlled via the bitcoin blockchain, using contracts.

Wallet—A collection of public and private keys, but may also refer to client software used to manage those keys and to make transactions on the bitcoin network.

Wallets don't actually store the money; they lock away access. The only way to get access to the money is by providing a password, as defined earlier, more commonly referred to as a "key." There are many types of wallets, both physical and digital, each with their own advantages and disadvantages. Typically, physical wallets offer more security features, but their drawback is that they take longer to access your cryptocurrency.

ABOUT THE AUTHOR

ERIC L. GUTHRIE, ESQ. is an experienced global trainer who has conducted legal, supervisory, teamwork, and diversity trainings in more than ten countries. Eric is the Director of Training for the Government Blockchain Association, and a Partner for the Cogent Law Group, a law firm with a FinTech, blockchain, and cryptocurrency practice. Eric is also a Certified Training Executive and a Certified Diversity Executive. Eric is bringing his training and writing skills to the cryptocurrency and blockchain world. He is partnering with numerous businesses and universities in several countries to host blockchain conferences. Learn more about Eric's services at BlockchainorDie.org or EricGuthrie.com.

NOTES

CHAPTER 1

[1] Nakamoto, Satoshi. "Bitcoin Open Source Implementation of a P2P Currency." *P2P Foundation* http://p2pfoundation.ning.com/forum/topics/bitcoin-open-source.

[2] Popper, Nathaniel. *Digital Gold: Bitcoin and the Inside Story of the Misfits and Millionaires Trying to Reinvent Money.* New York: HarperCollins, 2016.

[3] "Bitcoin" Bitcoin Wiki. https://en.bitcoin.it/wiki/Silk_Road (Accessed July 12, 2019)

[4] "Learn To Code for Free," *FreeCodeCamp.com,* https://www.freecodecamp.org/ (Accessed July 12, 2019)

[5] "Build for Developers," *GitHub.com,* https://github.com/ (Accessed July 26, 2019)

[6] Rosulek, Martin. "14 Bitcoin Quotes by Famous People" *Medium,* August 24, 2017. https://medium.com/@MartinRosulek/14-bitcoin-quotes-by-famous-people-6e7a1a009281

[7] "The Euro," *Europa.eu,* https://europa.eu/european-union/about-eu/euro_en (Accessed July 12, 2019)

[8] Economist Editorial Team. "Get Ready for the Phoenix," *The Economist,* January 9, 1988, Vol 306

[9] Reiff, Nathan. "Were There Cryptocurrencies Before Bitcoin?" *Investopia,* June 25, 2019. https://www.investopedia.com/tech/were-there-cryptocurrencies bitcoin/

[10] Ibid.

[11] Andrew, Paul. "Who is Nick Szabo?" *Coin Central*, April 15, 2018. https://coincentral.com/who-is-nick-szabo/

[12] Alvarez, Jose. "Who is Satoshi Nakamoto? We Look at the Possible Candidates, *Blockonomi,* March 18, 2019. https://blockonomi.com/who-is-satoshi-nakamoto/ https://blockonomi.com/who-is-satoshi-nakamoto/

[13] L.S. "Who is Satoshi Nakamoto?" *The Economist,* November 2, 2015. https://www.economist.com/the-economist-explains/2015/11/02/who-is-satoshi-nakamoto

[14] Nakamoto, Satoshi. "Bitcoin a Peer-to-Peer Electronic Cash System, Bitcoin.org, May 24, 2009. https://bitcoin.org/en/bitcoin-paper

[15] "Wikipedia: History of Bitcoin" Wikipedia Foundation, last modified July 10, 2019, 23:20, https://en.wikipedia.org/wiki/History_of_bitcoin

[16] Nakamoto, Satoshi. "Bitcoin a Peer-to-Peer Electronic Cash System" *Bitcoin.org* May 24, 2009. https://bitcoin.org/en/bitcoin-paper

[17] "Peer-to-Peer," *Decryptionary.* https://decryptionary.com/dictionary/peer-to-peer/ (Accessed July 12, 2019)

[18] "Peer-to-Peer," *Decryptionary.* https://decryptionary.com/dictionary/peer-to-peer/ (Accessed July 12, 2019)

[19] "Digital Signature," *Decryptionary.* https://decryptionary.com/dictionary/digital-signature/ (Accessed July 12, 2019)

[20] "Double Spending," Decryptionary. https://decryptionary.com/dictionary/double-spending/ (Accessed July 12, 2019)

[21] "Hash," *Decryptionary.* https://decryptionary.com/dictionary/hash-function/ (Accessed July 12, 2019)

[22] Nakamoto, Satoshi. "Bitcoin a Peer-to-Peer Electronic Cash System," *Bitcoin.org,* May 24, 2009. https://bitcoin.org/en/bitcoin-paper

[23] Nakamoto, Satoshi. "Bitcoin a Peer-to-Peer Electronic Cash System," *Bitcoin.org,* May 24, 2009. https://bitcoin.org/en/bitcoin-paper

[24] "Proof of Work," *Decryptionary.*
https://decryptionary.com/dictionary/proof-of-work/ (Accessed July 12, 2019)

[25] "Nodes," *Decryptionary,*
https://decryptionary.com/dictionary/node/ (Accessed July 12, 201

[26] "Public Address," *Decryptionary,*
https://decryptionary.com/dictionary/public-address/ (Accessed July 12, 2019)

[27] "Address," https://en.bitcoin.it/wiki/Address (Accessed July 12, 2019)

[28] "Altcoin," *Decryptionary,*
https://decryptionary.com/dictionary/altcoin/ (Accessed July 12, 2019)

[29] "Application Specific Integrated Circuit," *Decryptionary,*
https://decryptionary.com/dictionary/application-specific-integrated-circuit/ (Accessed July 12, 2019)

[30] "Blockchain," *Decryptionary,*
https://decryptionary.com/dictionary/blockchain/ (Accessed July 12, 2019)

[31] "Hard Fork," *Decryptionary,*
https://decryptionary.com/dictionary/hard-fork/ (Accessed July 12, 2019)

[32] "Soft Fork," *Decryptionary,*
https://decryptionary.com/dictionary/soft-fork/ (Accessed July 12, 2019)

[33] "Initial Coin Offering (ICO)," Decryptionary,
https://decryptionary.com/dictionary/initial-coin-offering/ (Accessed July 12, 2019)

[34] "Mining," *Decryptionary,*
https://decryptionary.com/dictionary/minting/ (Accessed July 12, 2019)

[35] "Private Key," *Decryptionary,*
https://decryptionary.com/dictionary/private-key/ (Accessed July 12, 2019)

[36] "Public Key," *Decryptionary*
https://decryptionary.com/dictionary/public-key/ (Accessed July 12, 2019)

[37] "Smart Contract," *Decryptionary,*
https://decryptionary.com/dictionary/smart-contract/ (Accessed July 12, 2019)

[38] "Wallet," *Decryptionary*
https://decryptionary.com/dictionary/wallet/ (Accessed July 12, 2019)

[39] Nakamoto, Satoshi. "Bitcoin a Peer-to-Peer Electronic Cash System," Bitcoin.org, May 24, 2009. https://bitcoin.org/en/bitcoin-paper (Accessed July 12, 2019) Nakamoto, Satoshi. "Bitcoin a Peer-to-Peer Electronic Cash System" *Bitcoin.org.,* May 24, 2009. https://bitcoin.org/en/bitcoin-paper (Accessed July 12, 2019)

[40] Nakamoto, "Bitcoin White Paper."

[41] Popper, Nathaniel. *Digital Gold: Bitcoin and the Inside Story of the Misfits and Millionaires Trying to Reinvent Money,* New York: HarperCollins, 2016.

[42] Popper, *Digital Gold.*

[43] Khatwani, Sudhir. "Five Different Crypto Wallets You Should Know About," *CoinSutr,* January 11, 2019, https://coinsutra.com/types-of-crypto-wallets/

[44] Khatwani, "Crypto Wallets."

[45] Khatwani, "Crypto Wallets."

[46] Khatwani, "Crypto Wallets."

[47] Khatwani, "Crypto Wallets."

[48] Multicurrency wallets

[49] Larson, Selena. "The Hacks That Left Us Exposed in 2017," *Money.cnn.com* December 20, 201, .https://money.cnn.com/2017/12/18/technology/biggest-cyberattacks-of-the-year/index.html

[50] Nakamoto, Satoshi. "Bitcoin a Peer-to-Peer Electronic Cash System," *Bitcoin.org,* https://bitcoin.org/en/bitcoin-paper

[51] Shrem, Charlie, "Bitcoin's Biggest Hack in History: 184.4Billion Bitcoin from Thin Air; Satoshi Hard Forks, Saves Bitcoin," *Harckernoon.com*, January 11, 2019, https://hackernoon.com/bitcoins-biggest-hack-in-history-184-4-ded46310d4ef?source=rss----3a8144eabfe3---4

52 Lee, Timothy, "A Brief History of Bitcoin Hacks and Frauds," Ars Technica, December 5, 2018, https://arstechnica.com/tech-policy/2017/12/a-brief-history-of-bitcoin-hacks-and-frauds/

[53] Zhao, Wolfie, "Bithumb $31 Million Crypto Exchange Hack: What We Know (and Don't), *CoinDesk.co,* June 21, 2018, https://www.coindesk.com/bithumb-exchanges-31-million-hack-know-dont-know/

[54] Nakamoto, "Bitcoin White Paper," Presented via email, May 24, 2009.

CHAPTER 2

[55] Shetty, Sameeepa. "James Altucher's 10 Predictions About Where Bitcoin and Cryptocurrencies are Headed, CNBC.com December 1, 2017 https://www.cnbc.com/2017/12/01/james-altuchers-bitcoin-predictions.html

[56] "CoinMarketCap," *CoinMarketCap.com.* https://coinmarketcap.com (Accessed July 12, 2019)

[57] "Methodology," *CoinMarketCap.com*, https://coinmarketcap.com (Accessed July 12, 2019)

[58] "Glossary," *CoinMarketCap.com.* https://coinmarketcap.com/glossary/ (Accessed July 12, 2019)

[59] "Top 100 Tokens By Market Capitalization," *CoinMarketCap.com, https://coinmarketcap.com/tokens/* (Accessed July 12, 2019)

[60] "Glossary," *CoinMarketCap.com.*

[61] "Top 100 Cryptocurrencies by Market Capitalization, *CoinMarketCap.com.* https://coinmarketcap.com (Accessed July 12, 2019)

[62] "Top 100 Cryptocurrencies," *CoinMarketCap.com.* https://coinmarketcap.com (Accessed July 12, 2019)

[63] Bitcoin.org, "Getting Started with Bitcoin," Bitcoin.org, https://bitcoin.org/en/ (Accessed July 12, 2019)

[64] CoinMarketCap

[65] "Ethereum, https://coinmarketcap.com/currencies/ethereum/ (Accessed July 12, 2019) https://ethereum.org

[66] "Total Payment Rails Don't Cut It," *Ripple.com* https://ripple.com, (Accessed July 12, 2019)

[67] "Best Money in the World," *BitcoinCash.org.* https://www.bitcoincash.org (Accessed July 12, 2019)

[68] Litecoin, "Money for the Internet Age," Litecoin.com. https://litecoin.com/en/ (Transcription of video on Litecoin.com - Accessed July 12, 2019)

[69] "EOS.io—Technical White Paper v.2," *Github.com,* March 16,2018, https://github.com/EOSIO/Documentation/blob/master/Tech nicalWhitePaper.md

[70] "Binance," *Binance.com, https://www.binance.com/en* (Accessed July 12, 2019)

[71] "Bitcoin SV," *BitcoinSV.io,* https://bitcoinsv.io (Accessed July 12, 2019)

[72] "Tether," *Tether.to,* https://tether.to (Accessed July 12, 2019)

[73] Stellar?

[74] "Top 100 Cryptocurrencies by Market Capitalization," *CoinMarketCap.com,* https://coinmarketcap.com (Accessed July 12, 2019)

CHAPTER 3

[75] Marr, Bernard. "Twenty Three Fascinating Bitcoin and Blockchain Quotes Everyone Should Read, *Forbes.com,* August 15, 2018. https://www.forbes.com/sites/bernardmarr/2018/08/15/23-fascinating-bitcoin-and-blockchain-quotes-everyone-should-read/#416a7c017e8a

[76] "What is the Difference Between Litecoin and Bitcoin?" *CoinDesk.com*, April 2, 2014, https://www.coindesk.com/information/comparing-litecoin-bitcoin.

[77] "Buy and Sell Cryptocurrency," *Coinbase.com*, https://www.coinbase.com (Accessed July 12, 2019)

[78] "Binance," *Binance.com*, https://www.binance.com/en (Accessed July 12, 2019)

[79] "S&P 500 Annual Total Return," *YCharts.com*, Accessed July 22, 2019, https://ycharts.com/indicators/sp_500_total_return_annual.

https://ycharts.com/indicators/sandp_500_total_return_annual (Accessed July 12, 2019)

[80] Bitcoin Exchange Guide News Team, "Bitcoin's Biggest Crashes Ranked—Volatility, Reasons & History?" *Bitcoin Exchange Guide*, September 22, 2017. https://bitcoinexchangeguide.com/biggest-bitcoin-crashes-ranked/.

[81] CoinMarketCap, https://coinmarketcap.com/currencies/ethereum/ (Accessed July 12, 2019)

[82] "Wikipedia: History of Bitcoin, *Wikipedia.com*, July 10, 2019. (Accessed July 12, 2019) https://en.wikipedia.org/wiki/History_of_bitcoin

[83] CoinMarketCap.

[84] Kenton, Will, "Bubble," *Investopia.com*, June 30, 2019, https://www.investopedia.com/terms/b/bubble.asp

[85] Hayes, Adam, "Dotcom Bubble, *Investopia.com*, June 25, 2019. https://www.investopedia.com/terms/d/dotcom-bubble.asp

[86] Seitz, Patrick, "Ten Big Tech Stocks That Climbed Back From The Dot-Com Crash," *Investors.com*, March 4, 2015. https://www.investors.com/news/technology/click/15-years-after-dot-com-crash-tale-of-stock-survivors/

CHAPTER 4

[87] Marinoff, Nick, "SEC Chairman: Cryptocurrencies Like Bitcoin are Not Securities, But Most ICOs Are," Nasdaq.com, June 7, 2018, https://www.nasdaq.com/article/sec-chairman-cryptocurrencies-like-bitcoin-are-not-securities-but-most-icos-are-cm975057.

[88] U.S. Commodities and Futures Trading Commission, "CFTC Statement on Self-Certification of Bitcoin Products by CME, CFE and Cantor Exchange, *U.S. Commodities and Futures Trading Commission.* December 1, 2017. https://www.cftc.gov/PressRoom/PressReleases/pr7654-17

[89] Marinoff, Nick, "SEC Chairman."

[90] El-Hindi, Jamal, "Application of FinCEN's Regulations to Virtual Currency Mining Operations," U.S. Treasury Financial Crimes Enforcement Network, January 30, 2014, https://www.fincen.gov/resources/statutes-regulations/administrative-rulings/application-fincens-regulations-virtual-0

[91] U.S. Office of Foreign Assets Controls.

[92] Aqui, Keith, "Notice 2014-21, *Internal Revenue Service,* https://www.irs.gov/pub/irs-drop/n-14-21.pdf

[93] Chen, James, "Commodity," *Investopia,* June 27, 2019, https://www.investopedia.com/terms/c/commodity.asp.

[94] Commodities and Futures Trading Commission.

[95] "Testimony of Chairman Timothy Massad Before the U.S. Senate Committee on Agriculture, Nutrition and Forestry," *U.S. Commodity Futures Trading Commission,* December 10, 2014. https://www.cftc.gov/PressRoom/SpeechesTestimony/opamassad-6

[96] Commodity Exchange Act. 7 U.S. Code § 1a (1936).

[97] Ibid.

[98] *Commodity Futures Trading Comm'n v. My Big Coin Pay, Inc.* *("My Big Coin"),* No. CV 18-10077-RWZ, 2018 WL 4621727 (D. Mass. Sept. 26, 2018).

[99] "About the SEC," *SEC.gov,* November 22, 2016,https://www.sec.gov/about.shtml

[100] "What We Do," *SEC.gov,* June 10, 2013, https://www.sec.gov/Article/whatwedo.html.

[101] Securities Act of 1933. 15 U.S. Code § 77a.

[102] Securities Exchange Act of 1934. 15 U.S. Code § 78a.

[103] Rooney, Kate. "SEC Chief Says Agency Won't Change Securities Laws to Cater to Cryptocurrencies, *CNBC,* June 6, 2018, https://www.cnbc.com/2018/06/06/sec-chairman-clayton-says-agency-wont-change-definition-of-a-security.html

[104] SEC v. Howey Co., 328 U.S. 293 (1946).

[105] "Framework for 'Investment Contract' Analysis of Digital Assets," US Securities and Exchange Commission. SEC.gov, April 3, 2019. https://www.sec.gov/corpfin/framework-investment-contract-analysis-digital-assets

[106] "SEC Issues Investigative Report Concluding DAO Tokens, a Digital Asset, Were Securities," *SEC.gov,,* July 25, 2017. https://www.sec.gov/news/press-release/2017-131

[107] Ibid.

[108] "Funds Raised in 2014," *ICOData.io,* https://www.icodata.io/stats/2014 (Accessed July 12, 2019)

[109] "Funds Raised in 2015," *ICOData.io,* https://www.icodata.io/stats/2015 (Accessed July 12, 2019)

[110] "Funds Raised in 2016," *ICOData.io,* https://www.icodata.io/stats/2016 (Accessed July 12, 2019)

[111] "Funds Raised in 2017," *ICOData.io,* https://www.icodata.io/stats/2017 (Accessed July 12, 2019)

[112] "Funds Raised in 2018," *ICOData.io,* https://www.icodata.io/stats/2018 (Accessed July 12, 2019)

[113] "Funds Raised in 2019," *ICOData.io,* https://www.icodata.io/stats/2019 (Accessed July 12, 2019)

[114] "SEC Exposes Two Initial Coin Offerings Purportedly Backed by Real Estate and Diamonds," *SEC.gov,* September 11, 2018, https://www.sec.gov/news/press-release/2017-185-0

[115] "SEC Emergency Action Halts ICO Scam," *SEC.gov,* December 4, 2017, https://www.sec.gov/news/press-release/2017-219.

[116] "SEC Emergency Action Halts ICO Scam," *U.S. Securities Exchange Commission,* December 4, 2017, https://www.sec.gov/news/press-release/2017-219

[117] "Company Halts ICO After SEC Raises Registration Concerns," *U.S. Securities Exchange Commission, December 11, 2017,* https://www.sec.gov/news/press-release/2017-227.

[118] "SEC Charges Digital Asset Hedge Fund Manager With Misrepresentations and Registration Failures," *U.S. Securities and Exchange Commission,* September 29, 2017, https://www.sec.gov/news/press-release/2018-186.

[119] "SEC Charges Bitcoin—Funded Securities Dealer and CEO," *U.S. Securities and Exchange Commission,* September 27, 2018, https://www.sec.gov/news/press-release/2018-218.

[120] "SEC Charger ICO Superstore and Owners With Operating As Unregistered Broker-Dealers," *U.S. Securities and Exchange Commission,* September 11, 2018, https://www.sec.gov/news/press-release/2018-185.

[121] Kenton, Will, "Who is Bernie Madoff," *Investopia.com,* May 4, 2019, https://www.investopedia.com/terms/b/bernard-madoff.asp.

[122] "SEC Charges Texas Man With Running Bitcoin-Denominated Ponzi Scheme," *U.S. Securities and Exchange Commission,* July 23, 2013, https://www.sec.gov/news/press-release/2013-132.

[123] "Framework for "Investment Contract" Analysis of Digital Assets," *U.S. Securities and Exchange Commission, April 3, 2019,* https://www.sec.gov/corpfin/framework-investment-contract-analysis-digital-assets.

[124] U.S. Securities and Exchange Commission, "Statement on Cryptocurrencies and Initial Coin Offerings," SEC.gov. December 11, 2017, https://www.sec.gov/news/public-statement/statement-clayton-2017-12-11.

[125] Aqui, Keith, "Notice 2014-21" Internal Revenue Service. https://www.irs.gov/pub/irs-drop/n-14-21.pdf (Accessed July 12, 2019)

[126] Aqui, Keith, "IRS."

[127] Aqui, Keith, "IRS."

[128] Brito, Jeffrey, "Hot Takes," *Coin Center,* June 6, 2017, https://coincenter.org/link/the-congressional-blockchain-caucus-co-chairs-asked-the-irs-for-better-guidance-on-digital-currency-taxation.

[129] Higgins, Stan, "U.S. Congressional Group Calls on IRS to Clarify Bitcoin Tax Guidance," CoinDesk, July 7, 2017, https://www.coindesk.com/us-congressional-group-calls-on-irs-to-clarify-bitcoin-tax-guidance.

[130] H.R. 2144: Token Taxonomy Act of 2019.

[131] FINCEN to Regulate Know Your Customer (KYC) and Anti-Money Laundering (AML).

[132] "Office of Foreign Assets Control- Sanctions and Program Information," *Treasury.gov,* April 5, 2019, https://www.treasury.gov/resource-center/sanctions/Pages/default.aspx.

[133] "OFAC FAQs: Sanctions Compliance," *Treasury.gov.* July 2, 2019 https://www.treasury.gov/resource-center/faqs/sanctions/pages/faq_compliance.aspx

[134] OFAC FAQs.

CHAPTER 5

[135] Rosenfeld, Everett, "Larry Summers: Overwhelmingly Likely This Will Change Finance Forever," *CNBC,* May 4, 2016, https://www.cnbc.com/2016/05/04/larry-summers-overwhelmingly-likely-this-will-change-finance-forever.html.

[136] Advanced Research Projects Agency Network (ARPANET), invented in 1983.

[137] Bloomberg, "Blockchain Is Pumping New Life Into Old-School Companies Like IBM and Visa," *Fortune Magazine, December 26, 2017.* https://fortune.com/2017/12/26/blockchain-tech-companies-ibm/.

[138] Hyperledger, "About Hyperledger," Hyperledger.com, https://www.hyperledger.org/about (Accessed July 19, 2019)

[139] "About," *Bitcoinmagazine.com,* https://bitcoinmagazine.com/about (Accessed July 12, 2019)

[140] Buterin, Vitalik. "Ethereum White Paper: A Next Generation Smart Contract & Decentralized Application Platform," *Github.com,* July 20, 2015.

[141] Ibid.

[142] Ibid.

[143] Ibid.

[144] Ibid.

[145] Veness, Chris, "SHA-256 Cryptographic Hash Algorithm," *Movabletye.co,* https://www.movable-type.co.uk/scripts/sha256.html (Accessed July 12, 2019)

[146] Lamport, Leslie; Shostak, Robert and Pease, Marshall, "The Byzantine Generals Problem," *SRI International,* November 1981.

[147] Ibid.

[148] "Consensus-Decision-Making, *Wikipedia.* July 8, 2019. https://en.wikipedia.org/wiki/Consensus_decision-making.

[149] Saini, Vaibhav. "ConsensusPedia: An Encyclopedia of Thirty Plus Consensus Algorithms, *Hacknoon.com* June 26, 2018. https://hackernoon.com/consensuspedia-an-encyclopedia-of-29-consensus-algorithms-e9c4b4b7d08f

[150] Ibid.

[151] Buterin, Vitalik, "Ethereum White Paper: A Next Generation Smart Contract & Decentralized Application Platform, *Github.com,* July 20, 2015

[152] Buterin, Vitalik, "Ethereum White Paper."

[153] Sai, "About, https://sia.tech/about (Accessed July 12, 2019)

[154] Ibid.

[155] Buterin, Vitalik, "Ethereum White Paper."

[156] Olarinoye, David, "What is a Decentralized Autonomous Organization (DAO)," *InvestInBlockchain.com,* July 3, 2018. https://www.investinblockchain.com/decentralized-autonomous-organization-dao/

[157] "Wikipedia: Organization," *Wikipedia.com,* June 7, 2019, https://en.wikipedia.org/wiki/Organization.

[158] Buterin, Vitalik, "Ethereum White Paper: A Next Generation Smart Contract & Decentralized Application Platform," *Github.com,* July 20, 2015.

[159] "Statistics," State of the DApps, https://www.stateofthedapps.com/statistics (Accessed July 23, 2019)

[160] Sherter, Alain, "Uber Hacked, Data for 57 Million People Exposed," *CBS News,* November 22, 2019, https://www.cbsnews.com/news/uber-hacked-data-for-57-million-people-exposed/.

[161] FairRide, "Ride Sharing on the Blockchain: The Future of Ride Sharing is Here," *Ridecoin, https://www.fairride.com* (Accessed July 12, 2019)

[162] La`Zooz, http://lazooz.org (Accessed July 12, 2019)

[163] BeeGlobal (now BeeNest), BeeNest, https://www.beenest.com (Accessed July 23, 2019)

[164] "CryptoBNB," *Cryptobnb.io,* https://cryptobnb.io (Accessed July 12, 2019)

CHAPTER 6

[165] Belmont, Clos, "Bitcoin Billionaire: Myth or Reality. How to Become Rich with Cryptocurrency Investment?" *Medium,* April 2, 2018, https://medium.com/clos-belmont/bitcoin-billionaire-myth-or-reality-how-to-become-rich-with-cryptocurrency-investment-3fc20cb33138

[166] IBM Food Trust Solution.

[167] "The Challenges of Storing Health Information Records," *Southeastern Oklahoma State University,* June 10, 2016, https://online.se.edu/articles/mba/the-challenges-of-storing-health-information-records.aspx.

[168] Schulte, Fred, "Fraud and Billing Mistakes Cost Medicare—and Taxpayers—Tens of Billions Last Year," *KHN.org,* July 19, 2017, https://khn.org/news/fraud-and-billing-mistakes-cost-medicare-and-taxpayers-tens-of-billions-last-year/.

[169] Herz, J.C., "Medicare Scammers Steal $60 Billion a Year," *Wired* March 7, 2016, https://www.wired.com/2016/03/john-mininno-medicare/.

[170] McGee, Marianne Kolbasuk, "$115 Million Settlement in Massive Anthem Breach Case, *DataBreachToday.com.* June 23, 2017. https://www.databreachtoday.com/115-million-settlement-in-massive-anthem-breach-case-a-10047

[171] Protonus, "Breached Patient Records Tripled in 2018 vs 2017, as Health Data Security Challenges Worry/" *Protenus,* February 12, 2019, https://www.protenus.com/press/press-release/breached-patient-records-tripled-in-2018-vs-2017-as-health-data-security-challenges-worsen.

[172] "State of the DApps," *StateoftheDapps.com,* https://www.stateofthedapps.com/rankings/category/health (Accessed July 23, 2019)

[173] Bradford, Contel, "Seven Infamous Cloud Security Breaches," *Blog.StorageCraft.com,* July 25, 2017, https://blog.storagecraft.com/7-infamous-cloud-security-breaches/.

[174] "Sharding, *Decryptionary,* https://decryptionary.com/dictionary/sharding/ (Accessed July 12, 2019)

[175] Storj Labs, Inc. "Storj: A Decentralized Cloud Storage Network Framework," October 30, 2018, https://storj.io/storj.pdf.

[176] Durant, Elizabeth; Trachy, Alison, "Digital Diploma Debuts at MIT," October 17, 2017, http://news.mit.edu/2017/mit-debuts-secure-digital-diploma-using-bitcoin-blockchain-technology-1017.

[177] Blockcerts, "Introduction," https://www.blockcerts.org/guide/ (Accessed July 12, 2019)

[178] Stein, Samantha, "Blockchain Engineers are in Demand," *TechCrunch,* https://techcrunch.com/2018/02/14/blockchain-engineers-are-in-demand/ (Accessed July 12, 2019)

[179] KPMG, "Safaricom: Ten Years of 'True Earnings,'" *Safaricom,* October 2016, https://www.safaricom.co.ke/images/Downloads/Resources_Downloads/True_Value_Booklet_Final.pdf.

[180] Monks, Kieron, "M-Pesa: Kenya's Mobile Money Success Story Turns 10," *CNN,* February 24, 2017. https://www.cnn.com/2017/02/21/africa/mpesa-10th-anniversary/index.html

[181] "M-Pesa," *Wikipedia.* https://en.wikipedia.org/wiki/M-Pesa (Accessed July 12, 2019)

[182] "Who Do You Want to Be?" *SteamRole.org,* https://steamrole.org/#home (Accessed July 23, 2019)

[183] "Complete Skill Roadmaps and Get Paid for Learning," SteamRole https://steamrole.org/#home (Accessed July 23, 2019)

[184] "The Power of Blockchain in Pursuit of Equal Rights," *LGBT Token,* https://lgbt-token.org/ (Accessed July 23, 2019)

[185] "About $GUAP," *GUAP,* https://guapcoin.com/newsite/about-us/ (Accessed 12, 2019)

[186] Ibid.

[187] Uulala, "The Blockchain Based Latino FinTech Company to Keynote AlPfa.org Cryptocurrency Education Event," *AuthorityPressWire.com,* March 7, 2018, https://authoritypresswire.com/uulala-blockchain-based-latino-fintech-company-keynote-alpfa-org-cryptocurrency-education-event/.

[188] Brumley, Jeff, "Is there a Market for a Christian Cryptocurrency? Christ Coin Aims to Find Out," *Baptist News*, November 3, 2017. https://baptistnews.com/article/market-christian-cryptocurrency-christ-coin-aims-find/

[189] B, Justin, "The List of Countries Where Bitcoin is Illegal or Restricted," Cryptovibes.com, December 18, 2018, https://www.cryptovibes.com/knowledge/the-list-of-countries-where-bitcoin-is-illegal-or-restricted/.

[190] DeSilva, Matthew, "Wyoming Goes All In On Blockchain, Cryptocurrency Bills," *ETHNews*, March 8, 2018, https://www.ethnews.com/wyoming-legislature-goes-all-in-on-blockchain-cryptocurrency-bills.

[191] Morton, Heather, "Blockchain State Legislation," *National Conference of State Legislatures*, March 28, 2019, http://www.ncsl.org/research/financial-services-and-commerce/the-fundamentals-of-risk-management-and-insurance-viewed-through-the-lens-of-emerging-technology-webinar.aspx.

[192] Erb, Kelly Phillips, "Ohio Becomes the First State to Allow Taxpayers to Pay Tax Bills Using Cryptocurrency," *Forbes.com*, November 26, 2018, https://www.forbes.com/sites/kellyphillipserb/2018/11/26/ohio-becomes-the-first-state-to-allow-taxpayers-to-pay-tax-bills-using-cryptocurrency/#6e9e41eb6b04.

[193] News Staff, "West Virginia Not Planning to Expand Use of Blockchain Voting, *Government Technology*, November 9, 2018. https://www.govtech.com/products/West-Virginia-Not-Planning-to-Expand-Use-of-Blockchain-Voting.html

CHAPTER 7

[194] Ulc, Lucia, "Why Usability and Design Directly Impact the ROI You Get From DAM," *Brandfolder*. https://brandfolder.com/blog/usability-dam/ (Accessed July 12, 2019)

[195] About," https://brave.com/about/ (Accessed July 12, 2019)

[196] CryptoKitties https://www.cryptokitties.co/ (July 12, 2019)

[197] The CryptoKitties Team, "CryptoKitties: Collectible and Breedable Cats Empowered by Blockchain Technology," *CryptoKitties.io,* (Accessed July 12, 2019) https://drive.google.com/file/d/1soo-eAaJHzhw_XhFGMJp3VNcQoM43byS/view.

[198] Ibid.

[199] CryptoKitties, https://www.cryptokitties.co/ (July 12, 2019)

[200] "CryptoPunks," Larvalabs.com, https://www.larvalabs.com/cryptopunks (Accessed July 21, 2019)

[201] Cossons, Malcolm, "Exploring Blockchain: Is the Art World Ready for Consensus," *Christies.com,* July 19, 2018, https://www.christies.com/features/Blockchain-and-the-art-market-9318-3.aspx

[202] "About," *Choon.co,* https://choon.co/about (Accessed December 10, 2018)

[203] "About," *Choon.co,* https://choon.co/about (Accessed July 12, 2019)

[204] "Voise," *Voise.com,* https://www.voise.com (Accessed July 12, 2019)

[205] Ibid.

[206] "Voise," *CoinMarketCap.com,* https://coinmarketcap.com/currencies/voisecom/ (Accessed July 22, 2019)

[207] "Provenance," *Provenance.org,* https://www.provenance.org/case-studies/martine-jarlgaard (Accessed July 12, 2019)

[208] "Casinoblockchain," *Casinoblockchain.io,* https://casinosblockchain.io (Accessed July 12, 2019)

[209] "Viola," *Viola.io,* https://viola.ai (Accessed July 12, 2019)

[210] Ibid.

[211] "What is Ponder?" *Ponderapp.com,* https://ponderapp.co (Accessed July 22, 2019)

[212] Spank Chain https://spankchain.com/company/ (Accessed July 12, 2019)

[213] Soleimani, Ameen and Bentley de Vogelaere, "SpankChain: A Cryptoeconomic Powered Adult Entertainment Ecosystem Built on the Ethereum Network," *Spankchain.com,* October 23, 2017, https://spankchain.com/static/SpankChain%20Whitepaper%20(EN).pdf

[214] Francis, Jeff, "Strippers Now Sporting Bitcoin Tattoos for Tipping," *Bitcoinist.com,* February 26, 2018, https://bitcoinist.com/strippers-now-sporting-bitcoin-tattoos-tipping/

[215] Schroeder, Stan, "HTC's New Phone Exodus Will Embrace the Blockchain," *Mashable.com* May 16, 2018, https://mashable.com/2018/05/16/htc-exodus/.

[216] "Sirin Labs," Accessed July 22, 2019, https://sirinlabs.com/.

[217] Shop.sirinlabs.com.

[218] The Airline Guru, "Delta Airlines is Ending SkyMiles and Replacing it With Cryptocurrency," *TravelCodex.com,* April 3, 2018, https://www.travelcodex.com/delta-air-lines-is-ending-skymiles-and-replacing-it-with-cryptocurrency/.

CHAPTER 8

[219] "Blockchain Quotes," *BrainyQuote.com,* https://www.brainyquote.com/topics/blockchain (Accessed July 12, 2019)

[220] Wesley, Daniel, "Is Blockchain the Invisible Answer to Your Blockchain Needs?" *Forbes.com,* December 11, 2017, https://www.forbes.com/sites/forbestechcouncil/2017/12/11/is-blockchain-the-invisible-answer-to-your-business-needs/#59b7493c13d0

[221] "About," *Consensys.net,* Accessed July 12, 2019, https://consensys.net/about/.

[222] "The Brooklyn Project," *Thebkp.com,* Accessed July 12, 2019, https://thebkp.com/about.

[223] Meunier, Sebastian, "When Do You Need Blockchain Decision Models?" *Medium.com*, August 4, 2016, https://medium.com/@sbmeunier/when-do-you-need-blockchain-decision-models-a5c40e7c9ba1.

[224] Ibid.

[225] Ibid.

[226] Ibid.

[227] General Services Agency, "Blockchain," *GSA.gov*, February 26, 2019, https://www.gsa.gov/technology/government-it-initiatives/emerging-citizen-technology/blockchain.

[228] "Contracting Data Analysis: Assessment of Government-wide Trends," GAO.gov, https://www.gao.gov/products/GAO-17-244SP (Accessed July 24, 2019)

[229] Criste, Laura; Snyder, Daniel; and Morris, Jodie, "Opportunities Show Growing Use of Blockchain," ITCON.com June 14, 2018, https://www.itcon-inc.com/content/opportunities-show-growing-agency-use-blockchain.

[230] Huillet, Marie, "U.S. Government Blockchain Spending Expected to Increase 1,000% Between 2017 – 2022: Study," *Cointelegraph.com, April 18, 2019,* https://cointelegraph.com/news/us-govt-blockchain-spending-expected-to-increase-1-000-between-2017-2022-study.

[231] "Five Quick Takeaways on Blockchain," *Fiscal.Treasury.gov.* March 20, 2018, https://fiscal.treasury.gov/fit/blog/five-quick-takeways-on-blockchain.html.

[232] Guthrie, Eric, *Diversify or Die*, Virginia: Better ME Better WE Publishing, 2016.

CHAPTER 9

[233] "Number of Sellers on Amazon Market Place," *MarketplacePulse.com,* https://www.marketplacepulse.com/amazon/number-of-sellers (Accessed July 12, 2019)

[234] By Airbnb, "Airbnb Hosts Share More Than Six Million Listings Around the World," *Airbnb.com,* March 1, 2019, https://press.airbnb.com/airbnb-hosts-share-more-than-six-million-listings-around-the-world/.

[235] Guta, Michael, "There are 168 Million Active Buyers on eBay Right Now," *Small Business Trends.* March 23, 2018.

[236] Uber Newsroom, "Company Info," *Uber,* https://www.uber.com/newsroom/company-info/ (Accessed July 12, 2019).

[237] "Upwork," *Wikipedia,* July 3, 2019, https://en.wikipedia.org/wiki/Upwork.

[238] Weyant, Jennifer, "Shark Tank Deal: Bundil Accepts $100,000 from Kevin O'Leary," *Business 2 Business Community.* October 27, 2018, https://www.business2community.com/entertainment/shark-tank-deal-bundil-accepts-100000-from-kevin-oleary-02134758.

[239] Bundil, https://enjoybundil.com/ (Accessed July 12, 2019)

[240] Coinseed, https://www.coinseed.co (Accessed July 12, 2019)

[241] Khatwani, Sudhir. "Nine Most Profitable Proof of Stake (POS) Cryptocurrencies," *CoinSultra,* March 10, 2018, https://coinsutra.com/proof-of-stake-cryptocurrencies/.

[242] Bitify, https://bitify.com/ (Accessed July 12, 2019)

[243] Purse, https://purse.io/shop (Accessed July 12, 2019)

[244] LBRY, https://lbry.com/ (Accessed July 12, 2019)

[245] Bitcoin Exchange Guide News Team, "Bitcoin's Biggest Crashes Ranked—Volatility, Reasons & History?" *Bitcoin Exchange Guide,* September 22, 2017, https://bitcoinexchangeguide.com/biggest-bitcoin-crashes-ranked/.

[246] Laxen, Matt, "Three Crypto Airdrops In Q3 2018 You Should Know About," *InvestinBlockchain.com,* July 10, 2018, https://www.investinblockchain.com/crypto-airdrops-q3-2018/.

[247] "Freelancing in America," *Edelman Intelligence,* Commissioned by Upwork and Freelancers Union, https://www.slideshare.net/upwork/freelancing-in-america-2017/1.

²⁴⁸ Coinbucks, http://coinbucks.io/ (Accessed July 12, 2019)

CHAPTER 10

²⁴⁹ Carmondy, Timothy, "Money 3.0: How Bitcoins May Change the Global Economy," *National Geographic,* October 15, 2013, https://news.nationalgeographic.com/news/2013/10/131014-bitcoins-silk-road-virtual-currencies-internet-money/

²⁵⁰ Budiman, Abby and Phillip Connor, "Migrants from Latin America and the Caribbean Sent a Record Amount of Money to Their Home Countries in 2016," *Fact Tank,* January 23, 2018, https://www.pewresearch.org/fact-tank/2018/01/23/migrants-from-latin-america-and-the-caribbean-sent-a-record-amount-of-money-to-their-home-countries-in-2016/.

²⁵¹ Pew Research Center, "Global Attitudes and Trends," *PewResearch.org,* April 3, 2019, https://www.pewresearch.org/global/interactives/remittance-flows-by-country/.

²⁵² Duffin, Aliesha, "ConsenSys, MakerDAO, and Dether, Launch Bifrost, a Project to Help Aid Groups in Conflict Zones," *Cryptoslate,* May 30, 2018, https://cryptoslate.com/consensys-makerdao-and-dether-launch-bifrost-a-project-to-help-aid-groups-in-conflict-zones/.

²⁵³ "Financial Inclusion on the Rise, But Gap Remain, Global Fidex Data Shows," WorldBank.org, April 19, 2018, https://www.worldbank.org/en/news/press-release/2018/04/19/financial-inclusion-on-the-rise-but-gaps-remain-global-findex-database-shows.

²⁵⁴ Torbati, Yeganeh, "Caribbean Counties Caught in the Crossfire of U.S. Crackdown on Illicit Money Flow," *Reuters,* July 12, 2016, https://www.reuters.com/investigates/special-report/usa-banking-caribbean/.

[255] Nel de Koker, Jeanne, "Why are Women in Developing Economies Excluded From Banking?" WeForum.org, September 16, 2015, https://www.weforum.org/agenda/2015/09/why-are-women-in-developing-economies-excluded-from-banking/.

[256] Demirgüç-Kunt, Asli, Leora Klapper, Dorothe Singer, Saniya Ansar, and Jake Hess, 2018, *The Global Findex Database 2017: Measuring Financial Inclusion and the Fintech Revolution.* Washington, DC: World Bank, doi:10.1596/978-1-4648-1259-0. License: Creative Commons Attribution CC BY 3.0 IGO

[257] According to the *International Federation of the Red Cross and Red Crescent Societies (IFRC) Report*: "Impact of the Regulatory Barriers to Providing Emergency and Transitional Shelter After Disasters, Country Case Study: Haiti."

[258] *Reuters,* "African Startups Bet on Blockchain to Tackle Land Fraud," *Reuters,* February 16, 2018, https://www.reuters.com/article/us-africa-landrights-blockchain/african-startups-bet-on-blockchain-to-tackle-land-fraud-idUSKCN1G00YK.

[259] Miller, Charles H. "Blockchain Land Records: 6 Countries That Are Testing the Technology As We Speak," *Wallet Weekly*, November 7, 2017, https://www.walletweekly.com/blockchain-land-records/.

[260] Mullings, David P.A. "Use Blockchain for Land Titling in Jamaica," *Medium,* January 26, 2018, https://medium.com/@davidmullings/use-blockchain-for-land-titling-in-jamaica-f45e6b947730.

[261] Campisi, Jennifer and White, Jaquetta, "Finally, Eleven Months After Maria, Power is Restored to Puerto Rico—Except for 25 Customers," *CNN,* August 7, 2018, https://www.cnn.com/2018/08/07/us/puerto-rico-maria-power-restored-wxc-trnd/index.html.

[262] "Our Mission," *Brooklyn.energy,* https://www.brooklyn.energy/ (Accessed July 24, 2019)

[263] Hande, Harish; Rajagopal, Surabhi; and Mundkur, Víkshut, "Energy for the Poor: Building an Ecosystem for Decentralized Renewable Energy," *Brookings,* https://www.brookings.edu/wp-content/uploads/2015/01/renewable-energy ch7.pdf (Accessed July 21, 2018)

[264] ImpactPPA, https://www.impactppa.com/ (Accessed July 12, 2019)

[265] Bates, Dan, "ImpactPPA: The World's Decentralized Energy Platform," *ImpactPPA,* March 2018, https://www.impactppa.com/.

[266] The Water Project, "A Year Later: Imbiakalo Community," TheWaterProject.org, December 2017, https://thewaterproject.org/community/interest_story/a-year-later-imbiakalo-community.

[267] The Water Project, "A Year Later: Kwambiha Community, *TheWaterProject.org,* November 2017. https://thewaterproject.org/community/interest_story/a-year-later-kwambiha-community

[268] The Water Project, "A Year Later: Kiluta Sand Dam, *TheWaterProject.org*, October 2017, https://thewaterproject.org/community/interest_story/a-year-later-kiluta-sand-dam.

[269] Hern, Alex, "It's Bobsleigh Time: Jamaican Team Raises $25,000 in Dogecoin," *The Guardian,* January 20, 2014, https://www.theguardian.com/technology/2014/jan/20/jamaican-bobsled-team-raises-dogecoin-winter-olympics.

[270] *Dache, Gerard,* "UN World Food Program Implements Blockchain and Realizes Massive Savings, GBAGlobal.org, May 13, 2018, https://www.gbaglobal.org/un-world-food-program-implements-blockchain-and-realizes-massive-cost-savings/.

[271] Mulligan, Cathy, "Blockchain and Sustainable Growth," *UNChronicle.org,* December 2018, https://unchronicle.un.org/article/blockchain-and-sustainable-growth.

[272] Ibid.

INDEX